AF678886

Sustainable Food Procurement

The book examines sustainable food procurement policy and practice in the European Union and beyond, exploring the extent to which sustainability objectives have been achieved and evaluating the new developments taking place at both EU and national levels.

While there is a growing recognition that public authorities can use public procurement as a policy tool to pursue multiple environmental, health and socio-economic objectives, contracting authorities still face many challenges. This volume investigates the scope for pursuing sustainable objectives in public procurement of food and catering services, examining different regulatory contexts and organisational models to answer the overall question of how to integrate sustainability concerns into the various phases of public food procurement processes. Contributions in the book examine the policy and legal procurement framework and practices for sustainable public catering in three EU Member States: Italy, France and Spain. There is a comparative survey of the Baltic Region, including Denmark, Estonia, Finland, Poland and Russia, and moving beyond the EU, there is examination of the UK and Brazil, as well as a cross country comparison of the UK with Denmark and Sweden. Drawing on the expertise of an interdisciplinary and intersectoral team of contributors allows the book to benefit from the insights of different disciplines, including business sciences, anthropology and law. Tapping into the global discussion on public food procurement as a means to achieve multiple social and environmental goals, this work will stimulate readers looking for new creative ways to create value through public food purchasing.

This book will be of great interest to students, researchers, policymakers and public- and private-sector representatives interested in public procurement, food policy and law, sustainable food sourcing and supply chain management.

Mark Stein is a research associate at the University of Salford, UK, where he completed his PhD on sustainable food procurement. He has decades of experience working in municipal economic development, where he supported small business development, secured external funding, promoted e-commerce and developed sustainable procurement policies.

Maurizio Mariani is the director of the international platform Eating City. He plays an active role in international research projects that aim to promote sustainable public food services. He has co-authored several books, such as *La Ville Qui Mange* (2013) and *City Food Policies: Securing our daily bread in an Urbanizing World* (2015).

Roberto Caranta is a professor of administrative law at the University of Turin, Italy, and coordinator of Sustainability and Procurement in International, European, and National Systems (SAPIENS) International Training Network. He is the author and editor of numerous books, including *European Public Procurement: Commentary to Directive 2014/24/EU* (2021).

Yiannis Polychronakis is the Head of Supply Chains, Procurement and Project Management in Salford University, UK. Yiannis is also an engineer with extensive industrial experience and has published his empirical work in numerous prestigious international journals and conference proceedings.

Routledge Studies in Food, Society and the Environment

Community Food Initiatives
A Critical Reparative Approach
Edited by Oona Morrow, Esther Veen, and Stefan Wahlen

Food Futures in Education and Society
Edited by Gurpinder Singh Lalli, Angela Turner, and Marion Rutland

The Soybean Through World History
Lessons for Sustainable Agrofood Systems
Matilda Baraibar Norberg and Lisa Deutsch

Urban Expansion and Food Security in New Zealand
The Collapse of Local Horticulture
Benjamin Felix Richardson

Evaluating Sustainable Food System Innovations
A Global Toolkit for Cities
Edited by Élodie Valette, Alison Blay-Palmer, Beatrice Intoppa, Amanda Di Battista, Ophélie Roudelle, and Géraldine Chaboud

How to Create a Sustainable Food Industry
A Practical Guide to Perfect Food
Melissa Barrett, Massimo Marino, Francesca Brkic, and Carlo Alberto Pratesi

Food Justice in American Cities
Stories of Health and Resilience
Sabine O'Hara

Sustainable Food Procurement
Legal, Social and Organisational Challenges
Edited by Mark Stein, Maurizio Mariani, Roberto Caranta and Yiannis Polychronakis

For more information about this series, please visit: www.routledge.com/Routledge-Studies-in-Food-Society-and-the-Environment/book-series/RSFSE

Sustainable Food Procurement

Legal, Social and Organisational Challenges

Edited by Mark Stein, Maurizio Mariani, Roberto Caranta and Yiannis Polychronakis

Routledge
Taylor & Francis Group
LONDON AND NEW YORK

earthscan
from Routledge

First published 2024
by Routledge
4 Park Square, Milton Park, Abingdon, Oxon OX14 4RN

and by Routledge
605 Third Avenue, New York, NY 10158

Routledge is an imprint of the Taylor & Francis Group, an informa business

© 2024 selection and editorial matter, Mark Stein, Maurizio Mariani, Roberto Caranta and Yiannis Polychronakis; individual chapters, the contributors

The right of Mark Stein, Maurizio Mariani, Roberto Caranta and Yiannis Polychronakis to be identified as the authors of the editorial material, and of the authors for their individual chapters, has been asserted in accordance with sections 77 and 78 of the Copyright, Designs and Patents Act 1988.

With the exception of Chapter 7, no part of this book may be reprinted or reproduced or utilised in any form or by any electronic, mechanical, or other means, now known or hereafter invented, including photocopying and recording, or in any information storage or retrieval system, without permission in writing from the publishers.

Chapter 7 of this book is freely available as a downloadable Open Access PDF at www.taylorfrancis.com under a Creative Commons Attribution-Non Commercial-No Derivatives (CC-BY-NC-ND) 4.0 license.

The Open Access version of Chapter 7 was funded by Università degli studi di Torino (University of Turin)

Trademark notice: Product or corporate names may be trademarks or registered trademarks, and are used only for identification and explanation without intent to infringe.

British Library Cataloguing-in-Publication Data
A catalogue record for this book is available from the British Library

ISBN: 978-1-032-48616-1 (hbk)
ISBN: 978-1-032-49297-1 (pbk)
ISBN: 978-1-003-39302-3 (ebk)

DOI: 10.4324/9781003393023

Typeset in Times New Roman
by Apex CoVantage, LLC

Dedication

Mark – To Carolyn

Maurizio – To the young minds shaping our future: May this book inspire you to nourish communities and foster sustainable change through the power of public food procurement.

Roberto – To the future of our kids

Yiannis – To Eleni, Emmanuel and to all prepared minds

Contents

Acknowledgements

Thanks to Leila Kentache for her invaluable assistance with the preparation of the manuscript. Thanks also for assistance received from Verity Stewart, Katie Stokes and Hannah Ferguson of Taylor & Francis.

Contributors

Urszula Ala-Karvia is a doctoral researcher at the University of Helsinki Ruralia Institute in Finland. Sustainability in public meals has been her research interest since 2019 when she contributed to the creation of the Sustainable Public Meal Toolkit through the Interreg Baltic Sea Region project StratKIT. Other research interests are urban-rural demographic changes and sustainable food chains.

Marta Andhov is an associate professor at the University of Copenhagen, Denmark. She spearheads significant research initiatives on sustainable public procurement, notably in sustainable food procurement. She co-hosts the Bestek Public Procurement Podcast and contributes to global procurement professionalisation through education and training.

Andrea Appolloni is an associate professor of sustainable procurement management at University of Rome Tor Vergata in Italy. He is a permanent visiting fellow at Cranfield University (UK). He is coordinator of the EU Project SAPIENS "Sustainability and Procurement in International, European, and National Systems" and coedits *Journal of Purchasing and Supply Management.*

Daniele Binci is an assistant professor in management at the University of Rome Tor Vergata, Italy, where he received his PhD in management. His research and teaching interests include innovation management, digital transformation and sustainability. He has published articles for international journals and presented papers at international conferences

Giulia Caddeo is a lawyer and researcher in public procurement at the Turin University Law Department. She has worked as an as intern for the City of Turin, developing green public procurement in food services. She has a deep understanding of European funded projects, especially in the fields of sustainability and food systems

Roberto Caranta is a professor of administrative law at the University of Turin, Italy, and coordinator of the Sustainability and Procurement in International, European, and National Systems (SAPIENS) International Training Network. He is the author and editor of numerous books, including *European Public Procurement: Commentary to Directive 2014/24/EU* (2021).

Wenjuan Cheng is a postdoctoral researcher at the University of Rome Tor Vergata, specialising in environmental economics and law. Her interdisciplinary research spans green public procurement, sustainable procurement and supply chain management, and biodiversity policy schemes.

Egidio Dansero is a full professor of political and economic geography at the University of Turin, Italy and Vice Rector for Sustainability and Development Cooperation. His main research interests are alternative food networks and urban food strategies. He coordinates the Italian Network on Local Food Policies (www.politichedelcibo.it), the university's food working group (https://reterus.it/cibo/) and the Turin food atlas (www.atlantedelcibo.it).

Chiara Falvo is a doctoral student in law at the University of Turin, Italy and an early-stage researcher at the SAPIENS Network, funded by the EU Horizon 2020 programme under the Marie Skłodowska-Curie grant agreement, no. 95669. Her research interests include public procurement and sustainable food systems.

Penelope Giosa is a senior lecturer in law at the University of Portsmouth, UK, and a qualified lawyer (not practising law) in Greece and Cyprus. Her research interests include sustainable food procurement, organic food legislation and the role of law in supporting agroecological transitions and ecological stewardship.

Robin Gourlay, OBE, worked for many years managing catering services within Scottish municipalities. He has produced guidance for the Scottish Government on food policy and public food procurement. He was a member of the Scottish Food Commission which drew up the policies now embodied in Scotland's Good Food Nation (Scotland) Act 2022.

Isabelle Hasquenoph is a lecturer (maître de conferences) in public law at Paris I Panthéon-Sorbonne University in France. She is the author of a thesis entitled *Public contracts and competition* (published in 2021). Her research focuses on public contract law, economic law and administrative law.

Ximena Lazo Vitoria is a professor of administrative law and coordinator of the Green Public Procurement Research Group, University of Alcala, Spain. She is the author and editor of several books and edited volumes, including *Green Public Procurement and Climate Change*, 2022 and *Local Produce and Public Procurement*, 2023.

Maurizio Mariani is the director of the international platform Eating City. He plays an active role in international research projects that aim to promote sustainable public food services. He has co-authored several books, such as *La Ville Qui Mange* (2013) and *City Food Policies: Securing our daily bread in an Urbanizing World* (2015).

Giorgia Masili is an assistant professor in the department of economics at Roma Tre University, Italy. She received her PhD in 2020 from the Carlo Bo University of Urbino, Italy. Her current research focuses on innovation management and sustainable business models for tangible cultural heritage.

Minna Mikkola developed her interest in food sustainability while teaching food hygiene and environmental protection to trainee chefs. Her PhD thesis in 2011 examined social dynamics for sustainable food systems. She has since worked in several Finnish and European research projects, at Helsinki University's Ruralia Institute – examining food and sustainability in public kitchens, farming, food processing, supermarkets and independent retail.

Sven Mikulic is a graduate research assistant at the Centre for Private Governance at the University of Copenhagen, Denmark. He specialises in EU public procurement law and policies, with a pronounced emphasis on investigating sustainability and digitalisation matters.

Line Rise Nielsen has a master's degree in political science from the University of Copenhagen, Denmark. She is the founder of Changing Food, a company specialising in sustainable change in professional kitchens. Changing Food is a partner in the European projects FoodSHIFT 2030 and Cocoreado, which focuses on public procurement.

José Pernas García is a professor of administrative law at the Universidade da Coruña, Spain, and a member of the Observatory of Environmental Policies, the Observatory of Public Procurement (obcp.es) of Spain and the executive committee of the Ibero-American Network of Public Procurement (redicop.com). He is also coordinator of the Knowledge Exchange Forum on Waste and Circular Economy (forocircular.com).

Yiannis Polychronakis is the head of supply chains, procurement and project management at Salford University, UK. Yiannis is also an engineer with extensive industrial experience and has published his empirical work in numerous prestigious international journals and conference proceedings.

Mark Stein is a research associate at the University of Salford, UK, where he completed his PhD on sustainable food procurement. He has decades of experience working in municipal economic development, where he supported small business development, secured external funding, promoted e-commerce and developed sustainable procurement policies.

Nadia Tecco is a sustainability officer and food manager at Turin University, Italy. Her research interests include public food procurement and circular economy. She is a contact person for the university's food working group (https://reterus.it/cibo/) and co-author of *Circular Economy for Food: Matter, Energy and Knowledge in a Circle*.

Introduction

Mark Stein, Maurizio Mariani, Roberto Caranta and Yiannis Polychronakis

Every bite of food we consume has a unique story of its journey and transformation. This book is a comprehensive guide that explores the complex process of procuring food and food services for public consumption. It sheds light on the often-overlooked aspect of our food systems – public food procurement – and covers the policies, practices and stakeholders involved in sourcing, purchasing and distributing food from farm to fork.

In the complex and ever-changing world of global markets, sourcing nutritious, sustainable and ethically produced food poses numerous challenges and opportunities. As a crucial link between agricultural production and consumption, public food procurement has the potential to bring about positive changes in the economic, social and environmental spheres. It embodies our society's values, reflecting our collective desire for fairness, sustainability and overall well-being.

Sustainable Food Procurement: Legal, Social and Organisational Challenges is a book that helps readers understand the complexities and nuances of food procurement for public institutions such as schools, universities, hospitals, elderly care, office canteens, the military and prisons.

It is now widely recognised that public procurement can be used as a policy tool to achieve various environmental, health and socio-economic goals, rather than just meeting the minimum requirements at the lowest cost.

The book discusses international frameworks, innovative models and case studies from different countries. It seeks to equip stakeholders with the knowledge and insights required to reshape and optimise public food procurement strategies.

The book aims to better understand the legal, social and organisational challenges associated with sustainable food procurement. Addressed to policy makers, procurement professionals, academics, students and concerned citizens, this book offers a window into the world of public food procurement, fostering a deeper understanding and inspiring informed action towards more sustainable and equitable food systems

Outline

Part I begins with country case studies examining various aspects of public food procurement. There is a focus on legislation, policies, practices and stakeholders

DOI: 10.4324/9781003393023-1

that influence how food is sourced, purchased, cooked and distributed on public plates. There are two case studies from the United Kingdom (Chapters 1 and 2) and two from Italy (Chapters 8 and 9). There are also general descriptions of sustainable food procurement law and policy in France (Chapter 5), Spain (Chapter 6) and Brazil (Chapter 7). Chapter 3 compares local food procurement policies in the UK, Denmark and Sweden. Chapter 4 compares public meal provision in five countries in the Baltic Sea Region: Denmark; Estonia; Finland; Poland and Russia. Chapter 10 considers how to design and measure the sustainability and social impacts of food service. Chapter 11 describes the European meal voucher and its evolution.

Part II examines future scenarios for sustainable procurement of food and catering. Chapter 12 considers the impact of the new European Union Farm to Fork Strategy. Chapter 13 considers world food systems and the potential for radical new approaches. Chapter 14 proposes a new paradigm for public food services.

Let's dive into the book chapters

Navigating through the realms of sustainable public food procurement presents an intricate, yet critical exploration of how public policies and practices intertwine with sustainability goals across various nations. This intricate weave encompasses various aspects from survey findings in English higher education, elucidating sustainability practices, to profound shifts in French public procurement law aiming at sustainable development across its multifaceted dimensions.

From England's support towards sustainability practices in higher education to the multilayered developments in French procurement laws, and from the diverging policies on school meals in the UK since 1979 to the distinctive approaches towards sourcing local food in Sweden, Denmark and Scotland, the picture painted is diverse yet bound by a common thread – sustainability. Further exploring the grounds of Italy, we witness a nation deeply rooted in sustainable public procurement, mandating Minimum Environmental Criteria in tenders, and deploying a holistic sustainable food procurement strategy. Meanwhile, Spain contributes to the tapestry by attempting to marry its legislative frameworks with sustainability, focusing on promoting sustainable public food procurement, and entwining measures from various thematic legislations into it.

The thread stretches over to Brazil, where the National School Feeding Programme emerges as a strategic shield against hunger and poverty while also promoting rural and sustainable development with family agriculture business and organic products. Insights from the University of Turin illuminate the role of universities in sustainable food procurement, revealing how a shift from a product-centric view to a "supply chain" logic can pivot the stakes of sustainability.

The evolution and impact of the meal voucher market over the past three decades in altering the dynamics of meals in public food services and commercial sectors brings forth an intriguing perspective on public food service sectors (Chapter 11). Moreover, diving into tools and indicators like Carbon Footprint, water footprint, and social return on investment (SROI), one uncovers the complex, interwoven impacts of our activities on environmental and social systems, particularly in the EU's food industry.

Legal framework of public procurement practice

Several chapters explore in detail the legal frameworks which shape the practice of public food procurement. Chapter 5 describes the evolution of public food procurement law and policy in France – particularly following the passing of the Egalim Law in 2018 which was a turning point in mandating sustainable food in public catering. Chapter 6 outlines Spanish legislation encouraging purchasing of local food for public kitchens – promulgated by both national government and regional administrations. It underlines the limitations in the Spanish legal framework and highlights the lack of detailed guidance for public procurers. Chapter 7 sets out the frameworks of law and policy for the school food system in Brazil which encourage purchasing from local family farmers and particularly from farmers with the most environmentally friendly practices such as organic agriculture or agroecology. Chapter 8 shows how the University of Turin worked within the Italian and EU legal frameworks to create a tender for vending machines which both promoted healthier food and was more environmentally friendly than before. Chapter 9 describes how the Italian government has created a set of Minimum Environmental Criteria for public food procurement which provides wide ranging and comprehensive sustainability guidance for public procurers.

Local food and short food chains

Public institutions play a vital role in promoting economic development and community resilience by choosing to purchase from local producers and suppliers. This helps boost small-scale farmers, local food businesses and regional economies, leading to a ripple effect of prosperity and stability. Chapter 1 considers the linkage between local food procurement and improving school food standards in the UK. Chapter 3 compares local food procurement in Sweden, Denmark and the UK, exploring significant differences between the three countries. The comparison of public kitchens in the Baltic countries in Chapter 4 shows how local food procurement is a widespread practice, but not found in all places. Chapter 5 refers to local food procurement initiatives in France. Chapter 6 shows how Spanish public procurement legislation encourages public kitchens to purchase local food. Chapter 7 describes how procurement legislation in Brazil provides detailed and specific guidance on procuring food from local family farmers. Chapter 9 shows how the Italian Minimum Environmental Criteria encouraged procurement from local suppliers and particularly from short food chains – especially organic food. Strong arguments are given as to the benefits of local purchasing in terms of lower environmental impacts and superior food quality. Chapter 10 describes the beneficial effects of procurement of high-quality local foods in the Scottish municipality of East Ayrshire. Chapter 14 describes public kitchens in the French department of

Dordogne as exemplary best practice, with sourcing of high-quality food which is both local and organic.

Public nutrition standards

Food in public kitchens should be nutritious and there should be public policies in place to ensure this. This topic is particularly discussed in five chapters. Chapter 1 describes the rise and decline of standards in UK school meal provision between 1980 and 2022. Policy in England has been particularly volatile. In Scotland and Wales there have been more consistent attempts to improve school food standards. Chapter 4 shows how all five Baltic Sea countries have public school food standards in place. Chapter 7 show how nutritionists in Brazil have a critical role in ensuring a high standard of school meals. Chapter 8 shows how the Turin University vending machine tender was designed to improve the nutritional standards of food provided by vending machines. Chapter 10 outlines the parameters of an ideal school canteen meal service, based on experiences in the Scottish municipality of East Ayrshire.

Who pays for public meals?

There are very considerable differences between countries as regards financial arrangements for providing children with school meals. Chapter 1 demonstrates the impact of financial stringency on school meal provision in the UK. It highlights the difference between England where most parents have to pay for school meals and Scotland, Wales and London where free primary school meals have been introduced. Chapter 4 describes public meal financing in five Baltic Sea countries. There is a range of different policies. On the one hand Finland provides free meals to all children. On the other hand, Denmark has only very limited provision of school meals. And Russia (St Petersburg) has a two-tier system. There is state subsidy for regulated social meals and a separate higher quality menu of choice for families with money. Chapter 5 describes payment arrangements for school meals in France, where some children receive free or subsidised meals. Chapter 7 describes the school meal system in Brazil, where all children are offered state funded meals.

A healthy planet

Public food procurement has considerable environmental implications. The Italian Minimum Environmental Criteria discussed in Chapter 9 amount to a comprehensive statement of the different environmental dimensions relating to public food procurement. The UK's Food for Life partnership has also produced a comprehensive environmental agenda for school catering (Chapter 1).

Organics and agroecology

Aspirations to purchase organic food for public kitchens are mentioned in several chapters. Chapter 1 shows how the percentage of organic food in UK public kitchens increased when school food standards were improving between 2014 and 2018 but fell steeply with deteriorating standards thereafter. Chapter 4 shows that percentages of organic food in public meals ranged from high levels in Denmark and Finland and lower levels in Poland and Estonia to zero in Russia (St Petersburg). Chapter 5 discusses moves to mandate a minimum percentage of organic food in France through the 2018 Egalim Law. Chapter 9 shows how the Italian Minimum Environmental Criteria provide a comprehensive argument for the virtues of organic food in public kitchens.

Organic agriculture may be mentioned alongside agroecology as more sustainable forms of agriculture. This is the case in Brazil, where public procurers are encouraged to favour these in procurement (Chapter 7).

As regards future scenarios, Chapter 12 refers to the EU Farm to Fork Strategy encouraging greater quantities of organic food in public kitchens. Chapter 13 posits a greater role for agroecology and organic food as the best way forward for a more sustainable food system. Chapter 14 presents public kitchens in the Dordogne in France as an exemplar of best practice with high levels of organic and local food.

Plant-based food

Several chapters stressed the importance of increasing plant-based foods in public procurement – on both nutritional and environmental grounds, while financial savings have also helped to finance an increase in organic food. On this development in UK school kitchens, see Chapter 1. On increasing plant-based food in English universities see Chapter 2. Chapter 4 provides an overview of five Baltic countries. Public kitchens in Denmark and Finland frequently have a high percentage of plant-based foods. The trend to increase plant-based food is still at an early stage in Poland and Estonia. In Russia (St. Petersburg) the public kitchens have made no moves to increase plant-based. Chapter 9 shows that the Italian Minimum Environmental Criteria encourage a higher percentage of plant-based foods.

Reducing GHG emissions

The Italian Minimum Environmental Criteria as described in Chapter 9 mandate a number of policy measures which are aimed at reducing GHG – such as avoiding purchases of vegetable oil produced through destruction of tropical rainforests. Chapter 10 describes the methodology for calculating the carbon footprint of public food procurement.

Fair food systems

Equity and ethics are foundational to building fair food systems. Food public procurement policies can address social inequalities by promoting fair labour practices and supporting marginalised communities.

The Italian Minimum Environmental Criteria set out in Chapter 9 make a strong case for procurement practices aimed at promoting decent working conditions within the farms providing the food for public kitchens. Three other chapters make reference to purchasing of Fairtrade products for public kitchens. These are Chapter 1, relating to school food in the UK, Chapter 2 relating to English universities, and Chapter 5, relating to France.

Changing kitchen practices

Chapter 1 refers to UK school kitchens changing their kitchen practices to improve the quality of school food – with more local and even some organic food. The kitchens moved to phase out prepared meals and ingredients, replacing them with cooking from fresh ingredients ("from scratch"). To make best use of local suppliers there was a move to seasonal menus. These changes produced savings, as did reductions in food waste and meat usage. Similar things were done by public kitchens in Denmark when they were converting to organic food. (See also Chapter 13.) Chapter 9 describing Italy's Minimum Environmental Criteria mandates a very similar set of cooking practices.

A very different tendency can be seen with the account of two central kitchens in Finland in Chapter 4. These kitchens were highly mechanised and efficient, producing very large volumes of prepared meals at low cost. One used cook-chill and the other a new process called Cook Cold.

Educating consumers

Policies aimed at improving the quality of school food may be accompanied by moves to develop food literacy – to educate children about healthy and sustainable food. This is sometimes referred to as a whole school food approach This may take such forms as school gardens, cooking lessons and visits to farms. In the UK this approach has been carried forward through the Food for Life programme which is described in Chapters 1 and 3. The comparison of local food procurement in the UK with Denmark and Sweden in Chapter 3 make clear that some Swedish public kitchens were also engaged in educating children in this way. One Swedish municipality called its food education project "Food for Life", showing that this activity was modelled on the UK project. Chapter 4 describes two schools in Poland where the wall separating the kitchen from the children's dining room was removed so that there could be dialogue between the cooks and the children. Chapter 10 is relevant here where it presents a case study of the social impacts of improvements in school food service.

Reducing food waste

Several chapters stress the importance of food waste reduction initiatives in public kitchens. This can be seen in the account of UK school kitchens in Chapter 1. Chapter 4 on the Baltic Sea Region shows that there were food waste reduction initiatives in Denmark, Estonia, Finland and Poland. Only Russia (St Petersburg) did not appear to have an initiative aimed at reducing food waste. Further references are made to reducing food waste in regard to France (Chapter 5), Spain (Chapter 6) and Italy (Chapter 9).

Measuring achievement

Chapter 10 describes methodology for measuring the achievements of public kitchens particularly in respect of carbon footprint (GHG emissions), Social Return on Investment – social, economic and environmental outcomes – and local economic multipliers. These concepts are explained in the chapter.

Future scenarios

In the second part of the book, a deep dive into the roots of EU public procurement regulation reveals a trajectory that traditionally vied to open domestic markets to EU-wide competition. Despite the entrenched orthodoxy of the internal market, environmental and social considerations have gradually permeated the EU public procurement regulatory framework. The transformative powers of the European Green Deal aim to break away from linear production and consumption models, veering towards a trajectory that harmoniously melds "how to buy" with "what to buy" under the umbrella of sustainable public procurement (SPP).

Through the lens of the Farm to Fork Strategy and its potential for mandatory SPP of food and catering services, the narrative unveils a global shift where governments, across the financial spectrum, are harnessing public food procurement initiatives as instruments to bolster sustainable development. Meanwhile, the chapter highlighting the unsustainability of the current industrialised global food system underlines the crucial role of agroecology, both as a set of practices and a social movement, emphasising the capability of public procurement to tackle the structural interlocks that sustain the industrial food systems in place.

Posing the question of why a paradigm shift in our food system and public procurement is imperative, the narrative underscores the overly industrialised nature of our food systems and its consequences. The need is to expand horizons, shifting focus from legal technicalities and concentrating on creating prerequisites that will later mould into the prerogatives of a new food system and a European strategy on sustainable public procurement of catering services. Engaging in an expansive dialogue about this, we delve deeper into varied case studies, such as the transformative school meals system in Dordogne, France, exploring the diverse yet integrally connected paths towards sustainable public food procurement across Europe.

Despite the growing popularity of plant-based, local, regional and place-based food paradigms, it is important to recognise that food literacy is still necessary. Procurement officers, food service managers, decision-makers and consumers, particularly the new generation, who are considered one of the most "vulnerable" groups in modern society, should be more aware about the importance of food literacy.

Implementing a multifaceted approach that combines education, family involvement, community engagement, policy advocacy and cultural appreciation such as the so-called "Whole Food Schools Approach" can effectively increase food literacy among the younger generation and promote healthier food habits.

Public food procurement emerges as a linchpin connecting agricultural production with consumption in the vast and bustling global markets. It operates at the intersection of economy, sustainability and public health, potentially driving transformative change across these landscapes.

The book, therefore, not only explores the variances and commonalities that define public food procurement across different nations but also strives to stitch together these disparate experiences and innovations, creating a narrative that transcends borders. It illuminates pathways, challenges and triumphs, crafting a journey that provides readers with insights and knowledge towards envisaging and shaping a future where public food procurement is universally sustainable, ethical and health-oriented.

Every morsel of food we consume carries a story of journey and transformation, and in these pages, we unravel the narratives woven into the tapestry of public food procurement.

Part I

National experiences

1 Rise and decline of standards in UK school meal provision, 1980–2022

Mark Stein and Yiannis Polychronakis

Introduction

This chapter describes how changes in UK policy on school meals have affected the quality of food provided to children over the period since the change of government in 1979. The chapter focuses particularly on England where government policy has seen a series of abrupt changes of direction. It shows how policy relating to England has increasingly diverged from policies adopted by the devolved governments in Scotland and Wales.

Overview of policy changes, 1980–2015

The policy changes at national level between 1980 and 2015 can be summarised as follows:

- Conservatives: Cutbacks in school food provision, 1984–1997.
- Labour: Improving school food, 2002–2010.
- Coalition: Cutbacks in school food spending, particularly with local authority caterers, 2010–2013.
- Coalition: Improves school food provision, 2013–2015 with School Food Plan and Universal Infant Free School Meals (UIFSM).

Thatcher: Cutbacks in school food provision, 1984–1997

In 1984 Mrs. Thatcher's government abolished the school food nutritional standards put in place in 1944 (Morgan & Sonnino, 2008, pp. 91–92). The statutory obligation to provide a school meal service for all children was abolished and some local authorities shut down their school kitchens, buying in sandwiches for poor children who continued to be eligible for free school meals. As part of a general drive to reduce costs, local authorities were obliged in 1984 to put their school catering services out to tender – compulsory competitive tendering. In most cases local authorities continued to provide school catering services but pay and ingredient costs were cut and meal quality declined. (Morgan, 2008, p. 1243).

DOI: 10.4324/9781003393023-3

Labour: Improving school food, 2002–2010

The Labour government elected in 1997 took steps to improve the quality of school food. During 2002–2009 stringent new school meal standards were introduced with strict nutritional requirements. Schools were encouraged to prepare food from fresh ingredients rather than ready meals (Morgan & Sonnino, 2008, pp. 93–95).

Between 2008 and 2010 the Labour government provided money to build new school kitchens and to establish training centres for school catering staff. Almost £500 million over three years of ring-fenced subsidy was provided to encourage children to eat school dinners (Department for Education, 2010).

Coalition: Cutbacks in school food, 2010–2013

The Conservative-Liberal Democrat Coalition government elected in May 2010 made very large cuts in local authority funding. The ring-fenced school meals grant was removed. The government encouraged schools to break away from local authority control and become academies directly funded by the Department for Education. New Free Schools were also set up independent of council control. Free Schools and Academies were exempted from nutritional standards (FFLP, 2010; Morgan, 2011; Nelson, 2012; Children's Society, 2012).

From 1 April 2013, almost all funding previously under local authority control – including that for school meals – was devolved to individual schools. Larger schools particularly were tempted to break their ties with local authority catering organisations and buy their catering service from a cheaper provider. (Previously a uniform local authority charge to all schools had meant that loss-making lunch services in small schools were cross-subsidised by larger schools – Dimbleby & Vincent, 2013, p. 106).

Coalition: Improves school food provision, 2013–2015

An abrupt policy reversal now took place. Late in 2012 the government had responded to criticism that allowing academies exemption from school food nutrition regulations was undermining child health, by appointing two celebrity chefs to carry out a review (Poulter, 2012). The review's findings were announced on 12 July 2013 – the School Food Plan.

Government policy had swung round to promoting the improvement of school catering services – building on the positive achievements of the previous government.

The government announced that, from September 2014 all state funded infant school children (i.e., those in reception, year 1 and year 2) would receive a free school lunch. It was anticipated that this would promote healthy diets and better academic performance and behaviour, particularly among the poorest pupils.

This involved a considerable increase in public spending – over £1 billion between 2014 and 2016, including £150 million of capital funding in 2014–2015 to improve kitchen and dining facilities. Schools would be given a uniform payment of £2.30 for every free meal provided.

Case study: Lancashire County Council

Lancashire County Council is one of the largest English local authorities, covering a population of 1.2 million.

The introduction of compulsory competitive tendering in 1989 led to cost cutting and impaired food quality. While contracts were awarded to the in-house Catering Direct Service Organisation (DSO) in 1989, and the Authority retained nutrition-based standards, there was a significant reduction in labour as more pre-prepared and processed food was procured. Spend on ingredients was reduced and less skill was required in preparation. The second round of tendering in 1994 saw fierce private sector competition and price rather than quality dominated contract award criteria. In 1994, all school catering contracts were transferred to BET Catering, along with all employees under Transfer of Undertakings Protection of Employment (TUPE) legislation. This brought increased pressure to reduce food and labour costs. Staff training was slashed.

In 1998, the contracts were awarded in-house and in 1999, the catering service was merged with cleaning and school crossing patrol Services to form Lancashire County Commercial Services (LCCS).

> The School Meals Service was at a low point. . . . Contracts provided no incentive for sales growth or innovation, and there was a lack of investment, poor front of house image and lack of promotional activity. . . . In 1999, the Authority, faced with difficult budget decisions, increased the selling price by an above inflation rise of 12p to £1.50 in Primary Schools. . . making Lancashire one of the most expensive. . . in England. A perception by parents of poor value for money, combined with falling rolls led to declining take up of meals and consequently many kitchens were operating well below capacity leading to poor productivity, cuts in staff hours, and resulting low morale. Many Headteachers and Governors were disillusioned with. . . the service.
>
> (Lancashire County Council, 2005)

In 2000 the government abolished compulsory competitive tendering. County Commercial Services now adopted a new strategy. This aimed to increase the numbers of children buying school meals by improving meal quality. Increased customer volume would make it possible to make further quality improvements without increasing prices.

The strategy comprised four main elements:

- Incentives: Higher sales enabled frozen selling prices for parents, lower costs recharged to schools, improved menus for customers, increased hours for employees and money for LCCS to invest in improvements.
- Innovation: New improved menus tailored for individual schools, renewed in autumn and summer to prevent menu fatigue, and backed by marketing.
- Involvement: Extensive market research of stakeholders and involving front line employees in promoting the service.

- Investment: Promoting healthy eating, improving dining rooms, introducing cashless payment systems and constructing kitchens in schools to replace transported meals.

This strategy increased take up of school meals from 31.2% of pupils in 2000/2001, to 44.5% in 2004/2005. Increased revenue allowed better quality menus with more variety while the selling price was frozen at £1.50 for six years. LCCS were able to make incremental improvements to the quality of food with each new menu launch.

These reforms laid the basis for a strong school catering service which has been able to continue down to the present. The school catering service has played an important economic role as a market for the county's agricultural produce (Lancashire County Council, 2022).

Development of the "Food for Life" concept, 2002–2006

In countries like Denmark or Sweden the principal measure of sustainable food procurement policy has been the percentage of organic food purchased. This has reflected pro-organic government policies and a high percentage of organic food in the consumer market. In the UK by comparison government policy has been lukewarm on organic and the percentage in the consumer market much lower.

The Soil Association is a long-established UK NGO which seeks to promote healthy and sustainable food, particularly organic food. In 2002, it decided to get involved in the moves to promote more healthy and sustainable food in public kitchens, particularly in schools (Pearce, 2006). The core of the Food for Life concept was that promotion of healthy and sustainable food in school kitchens and dining rooms would be interlinked with development of food education programmes for children inside schools. The children would learn about healthy and sustainable food by growing their own in school gardens, cooking lessons and visits to local farmers.

The Food for Life catering concept was first piloted on a large scale by Robin Gourlay, the Catering Manager in East Ayrshire in Scotland, who tried it out in 11 schools during 2004–2005. The goal was to purchase 30% organic and 50% local food, and that 75% of the food would be cooked using fresh ingredients rather than buying in pre-prepared meals. Food education would be scheduled alongside the transformation in the school kitchens. In August 2006, a Scottish Government study pronounced the East Ayrshire pilot a success:

> ingredient and administration costs have gone up modestly, though only within the range that many Local Authorities are already paying. . . . The approach has raised the quality of ingredients, . . . The social benefits felt by children, parents and teachers seem to be driven by the "freshness & healthiness" of the food supply, strongly linked to its "local" sources. "Organic" has been only a secondary driver, suggesting that social benefits can be achieved without added constraints on supply and cost from any organic requirement.
> (Bowden et al., 2006, pp. 4–5)

Food for Life Partnership national launch, early 2007

The Food for Life Partnership (FFLP) was an alliance of the Soil Association with three other NGOs – Focus on Food, Garden Organic and the Health Education Trust.

It received a very large grant – £17m over five years – from the Big Lottery Fund (Morgan, 2011). This enabled it to work with schools in every English region employing a total of around 80 staff. The core idea was to enable children to eat good food, learn where it comes from, how it is produced and how to grow and cook it themselves. FFLP also involved:

- Parents who came into school and share cooking skills.
- Teachers who integrated food into the curriculum.
- Head teachers who led the process.
- School catering staff who prepared the food from fresh organic and local ingredients.
- Local farmers and food producers.

FFLP was designed from the outset so that school caterers could win a hierarchy of awards and progress would be publicly recognised at each level, bronze, silver, and gold (see Table 1.1).

Initially the minimum organic percentage required for Gold was 30%. This reflected government support for increased public purchasing of organic food in the *English Organic Action Plan* (DEFRA, 2002, p. 12). Since 2010 there has been much less government support for organic food. Over the last 20 years there has been overall much less support for organic agriculture in the UK compared to other European countries (Mottershead & Maréchal, 2017, pp. 63–64). In 2011 the Soil Association reduced the minimum requirements to 15% organic for Gold and 5% for Silver.

The FFL approach involved a highly structured scoring system whereby school catering organisation were inspected annually by a visit from the Soil Association

Table 1.1 Food for Life Bronze, Silver and Gold Criteria

Bronze Award criteria	*Silver Award Criteria*	*Gold Award Criteria*
75% of dishes on menu freshly prepared	Range of local items	At least 15% of ingredients are certified organic or MSC
Seasonal menus	Range of certified organic items	At least 50% of ingredients local
Eggs cage free	Poultry eggs and pork are Freedom Food	Emphasis on animal welfare
Meat Farm Assured as minimum	Only sustainable fish	Increased Fairtrade and vegetarian food
Minimise additives and no hydrogenated fat	At least one Fairtrade product	

Source: FFLP, 2011; FFLP, 2013

inspector who visited kitchens and dining rooms and examined menus and a sample of purchasing records to verify organic and other accreditations and awarded points according to the criteria set out in the FFL Standards Handbook – which for schools was 47 pages long (Soil Association, 2019c). For example – to encourage purchasing of local and UK food, caterers were awarded 3 points per % of their spend on raw ingredients produced in their region and two points per % of spend over 59% spent on raw ingredients produced in the UK.

More organic food

Table 1.2 shows that the numbers of schools with FFL Silver and Gold accreditation rose considerably between January 2013 and January 2018, denoting an overall improvement in food quality, particularly increased usage of organic food. The percentage of schools with FFL Gold rose from 15.1% to 20.5%. The percentage of schools with FFL Silver rose from 24% to 46.6%. This reflects the mood of optimism in school catering in the years immediately after the School Food Plan, the financial boost resulting from UIFSM and the renewed emphasis on food quality with the publication of new food standards. During these years it appeared that school catering in the UK was moving in the same direction as other European countries which had made large increases in the usage of organic food.

UK school caterers endeavoured to improve the quality of school food in a cost neutral fashion. Quality improvements and better marketing increased sales of school meals and uptake of free meals. Moreover, costs were reduced through cooking from scratch, seasonal menus and reducing food waste and meat usage (School Food Matters, 2014; FFLP, 2021, p. 15). These were the same methods deployed by Danish catering managers to achieve cost neutral organic conversion (Mikkelsen & Sylvest, 2012; Sørensen et al., 2016).

Table 1.2 UK Local Authority School Catering – Growing Proportion of Silver and Gold, 2013–2018

	Bronze	*Silver*	*Gold*	*Total*
18th Jan 2013	1,518	599	376	2,493
% of total	60.9	24.0	15.1	100
"7th Feb 2015"	2,679	1726	572	4,977
% of total	53.8	34.7	11.5	100.0
"25 Feb 2016"	3,620	1,748	905	6,273
% of total	57.7	27.9	14.4	100.0
"31 Jan 2018"	1905	2,698	1,187	5,790
% of total	32.9	46.6	20.5	100.0

Source: Collated by authors from data published by Soil Association[1]

London survey: Improving standards, 2011–2018

Since 2011 the London mayor has funded the NGO Sustain to produce an annual report comparing sustainable and healthy food developments among the 33 London boroughs (Dalmeny et al., 2011; Compton, 2014; Parente, 2016; Guerlain, 2018; Davenport, 2019). Commitment to animal welfare in food procurement has risen from nine boroughs in 2011 holding a Good Egg award to 19 boroughs in 2019. The number of sustainable city food partnerships has risen from three boroughs in 2014 to eight in 2018 (Davenport, 2019, pp. 10–15). By 2019, 27 boroughs were taking actions to support sustainable fish. With Food for Life the number of boroughs which had at least Bronze for the majority of schools rose from 17 in 2011 to 23 by 2018. In 2018 three boroughs had reached the highest achievement level – a borough achieving Gold Food for Life for the majority of schools and for one or more additional sectors of catering under council control.

Deteriorating finance, 2016–2021

The 2015 general election was won outright by the Conservative Party. A crucial development was that post-2016 the government made substantial reductions to the school budget, whereas it had been largely protected from cuts under the 2010–2015 coalition government (Belfield & Sibieta, 2015).

The result was that established council catering organisations frequently found that schools were no longer buying their services – choosing cheaper alternatives, private catering contractors or in-house provision (APSE, 2020, pp. 16–17). Local authority caterers have consequently faced a constant battle to retain schools – knowing that if their organisation shrinks below a certain level it will cease to be viable. This is illustrated by a report from the City of Leeds school catering organisation (Leeds City Council, 2017, pp. 16–17).

In 2019 the London Borough of Greenwich decided that it needed to shut down its arm's length school catering organisation, GS Plus. The Chief Executive said that GS Plus faced a climate in which

> costs are increasing, but the company is experiencing a decline on income. This has left the company on the edge of insolvency. . . . Catering and school cleaning face significant pressures. Part of this is down to school funding cuts and the difficult decisions that governing bodies face on how to best spend their resources. . . . The bottom line is, as a supplier of school meals, the company relies on schools' finances and they are simply not buying them anymore. . . school budgets are absolutely on the floor.
>
> (Bull, 2019)

Schools desperate to save money were the more able to bring in cheaper private caterers because the government took no action to enforce the school food standards which had been put in place in 2014 and which local authority caterers generally continued to follow (Soil Association, 2019b, pp. 20–21). A catering contractor commented that

> In England there is no inspection of school meals. I think that is a big mistake. Caterers promise all sorts to schools when bidding for a contract – Food for Life and all that When they are in the school, quality goes down very quickly. Some inspection would result in better quality for the children.
>
> (Unpublished Interview, 25 March 2021)[2]

The 2021 House of Commons enquiry concluded that:

> It is essential that the Government sets buying standards for public procurement of food. . . standards are only effective if they are followed. . . . A responsibility for monitoring compliance should be emphasised to or placed on existing inspection bodies, such as Ofsted (for schools) and the Care Quality Commission (for NHS Trusts).
>
> (House of Commons Environment, 2021, p. 21)

UIFSM grant not uprated with inflation, 2014–2023

The payment per meal for Universal Infant Free Schools Meals was fixed at £2.30 when these were introduced in September 2014 and as the years went by the government did not update this payment to take account of inflation. It is scarcely surprising therefore that the quality of school food deteriorated in these years (Soil Association, 2019b).

The introduction of UIFSM had been an important initiative of the Coalition Government and very much championed by Nick Clegg, leader of the Liberal Democrats (Chiu, 2014). It lacked political support within the Conservative government which won an outright majority in 2015. In 2017 the prime minister, Theresa May, called a general election and her policy platform included abolition of UIFSM and replacement by a much cheaper school breakfast scheme (Bradshaw, 2017). This policy proved electorally unpopular and Theresa May failed to win a parliamentary majority and had to continue as a minority government. While unable to abolish UIFSM the government could refuse to uprate the UIFSM payment with inflation.

It was only on 14th July 2020 that the Education and Skills Funding Agency (EFSA) increased the payment to £2.34. There were further increases to £2.41 in June 2022 and £2.53 from September 2023. In real terms the payment has fallen very considerably at a time of rapid increases in food prices (Weale, 2022; Price, 2023).

London: Deteriorating food standards, 2018–2021

The 2019 *Good Food for London* report provides clear evidence of deteriorating school food standards among the London boroughs. With Food for Life the number of boroughs which had at least Bronze for the majority of schools had fallen to 19 compared to 23 in 2018 (Davenport, 2019).

In 2019 only one borough – Islington – had reached the highest achievement level – Gold Food for Life for the majority of schools and for one or more additional sectors of catering under council control, compared to three in 2018 (Davenport, 2019). The October 2020 Sustain report focused on London's food response to COVID-19. The report referred only briefly to public procurement and made no reference to Food for Life (Oliver-Larkin & Luck, 2020).

Closure of school catering organisations

A number of local authorities shut down their school catering organisations during the years 2018–2021. These included Bath and North East Somerset, Cambridgeshire, London Borough of Greenwich, Luton and Manchester. These were all catering organisations which had considerable achievements to their credit. Greenwich school catering was the scene of Jamie Oliver's televised experiment with improving school meals and had subsequently achieved FFL Gold. (Morgan & Sonnino, 2008, pp. 103–112). Cambridgeshire and Bath & North East Somerset had both worked extensively to source high-quality food for their school kitchens, the latter making use of an innovative dynamic purchasing system.

In February 2020, Manchester Fayre was praised as an exemplar of best practice in school catering – FFL Bronze, high-quality food, much locally sourced and excellent employment conditions (APSE, 2020, p. 41). Ten months later Manchester City Council announced its decision to dissolve Manchester Fayre on the grounds that so many Manchester schools had found alternative catering arrangements that it was no longer viable (Manchester City Council, 2020).

Collapse of Food for Life Gold, 2018–2022

In January 2018 there were eight English local authorities where all or most schools had Food for Life Gold (Soil Association, 2018). Our assessment is that by October 2021 this number had fallen to only one, Leicestershire. Six of these councils had gone down to FFL Silver – which meant that they bought up to 5% organic food: Cheshire East; Nottinghamshire and the London Boroughs of Barking and Dagenham, Tower Hamlets and Waltham Forest. Two had ceased to have FFL accreditation: Oldham and Surrey. And one catering organisation had been closed down – GS Plus (Greenwich).

Havering Council said in 2019 that uncertainty over food supplies created by Brexit was the reason why it decided to reduce buying organic food and take its FFL accreditation down to Bronze (Soil Association, 2019a).

The reasoning behind Tower Hamlets' decision to reduce organic food in school meals was set out in a committee report (Tower Hamlets Council, 2019). It was argued that the Council needed to make savings to eliminate the budget deficit without increasing the £2.30 meal price, which would cause more schools to defect to cheaper private caterers. Reducing organic food from 15% (FFL Gold) to 5% (FFL Silver) would save £342,000 per annum, or £0.08 per meal.

The October 2021 Soil Association report on Food for Life stated that annual spend on organic ingredients from sites serving a Silver or Gold menu was £12,480,867 during the year from 1 June 2020 to 31 May 2021 (FFLP, 2021). The Soil Association chose this period because schools were closed due to the pandemic during April and May 2020 and consequently no food was bought. This figure was 53.9% down on the £23.2m spent in 2018/2019 (Soil Association, 2021).

By mid-2023, rising ingredient costs had caused Leicestershire to cut back on organic food and pause its FFL accreditation, although it hoped to resume FFL accreditation at Silver.

National food strategy: Need to improve public food

On 27 June 2019, the government had commissioned Henry Dimbleby to conduct an independent review to help create the first National Food Strategy for 75 years. The second National Food Strategy report published in July 2021 recommended radical changes in the government's approach. It envisaged giving much higher priority to health, animal welfare, restoring the natural environment, combating climate change, and promoting resilience to global shocks (Dimbleby, 2021, p. 2). The government should work with existing certification bodies such as Food for Life to introduce a mandatory accreditation scheme for all public institutions, requiring them to achieve minimum standards for public food and encouraging them to aim higher. The Food Standards Agency should monitor the quality of public sector food and compliance with minimum standards.

England: School food standards compliance pilot

In November 2022, the Food Standards Agency and Department for Education (DfE) launched a year-long School Food Standards Compliance pilot across 18 participating local authorities in England (Food Standards Agency, 2022).

The pilot aimed to design and test a new approach in supporting schools to comply with the existing school food standards. Eighteen English local authorities agreed to participate. Their role would include suggestions as to how the school food standards could be improved and checks into the actual menus in use in schools. It was envisaged that where non-compliance with the school food standards was identified, local authorities would then work with schools to instigate supportive interventions to aid compliance with the standards.

Local authorities taking part this term have begun introducing some additional questions, as well as making observations related to the school food standards. This could include nutrition-focused questions and observations such as menu checks. These additional questions and observations would identify any potential non-compliance with the school food standards. Local Authorities would then work with schools to instigate supportive interventions to aid compliance with the standards. The pilot was intended to run in line with the full academic year. At the time of writing (August 2023) further news of the pilot project is expected.

London introduces free primary school meals

On 20 February 2023, the Mayor of London announced that there would be a one-off payment of £130m to ensure that all primary schoolchildren in the capital would receive free school meals for the 2023/2024 academic year (London, 2023). This would benefit around 270,000 primary school children and save families in London around £440 per child across the year. It would support families struggling with the spiralling cost of living. It also represents a welcome injection of public money into the school food system. In justifying this funding decision, the mayor cited an evidence briefing by the Food Foundation which documented the beneficial effects of free school meal policies (Food Foundation, 2022).

Scotland and Wales diverge from England

Over the last 25, years policies of the devolved governments in Scotland and Wales have increasingly diverged from those adopted in England.

Scotland

Scottish government policies have provided local authority school caterers with a much more favourable environment than in England.

- There has been a longstanding and sustained commitment to provision of high-quality food in school kitchens. This can be traced back to the report *Hungry for Success*, published in 2002 (Scottish Government, 2002).
- Almost all schools continue to have meals provided by the local authority catering organisations. There has been very little outsourcing.
- The Scottish government has enforced school nutritional standards as part of the school inspection regime provided by Education Scotland. Nutritional standards were tightened early in 2021 – requiring local authorities to reduce usage of processed meat in school meals and increase provision of vegetables.
- To promote better quality in public sector food the Scottish government has provided funding for a team of ten people at Soil Association Scotland – encouraging local authorities to adopt Food for Life Served Here. The number of Councils with FFL accreditation rose from ten in March 2019 to sixteen in 2021 – half of all Scottish Councils – and 17 by July 2023.
- The late 2021 decision by the Scottish government to introduce free meals for all children in primary schools has injected more public money into school catering organisations.
- At the present time the only part of the UK which has a current action plan to promote organic food is Scotland, which encourages public procurers to buy organic (Scottish Government, 2016, p. 11).

Some Scottish councils face very significant financial pressures, and this has affected the quality of their school meals. Concerns have been expressed that the Scottish government may not provide adequate funding for councils to provide universal primary school meals.

Wales

There are important similarities between Scotland and Wales regarding public food procurement policies. The Welsh government moved in late 2022 to introduce universal free primary school meals. Of 22 local authorities in Wales, 20 have retained in-council school catering services. Primary schools have stayed with the local authority provision, but many secondary schools use private caterers. Seventeen local authorities provide catering for most of their secondary schools.

Where private caterers do hold school food contracts in Wales, the standards are poor.

> There is one contractor I don't want to name who is coming into Wales. It is very noisy about its sustainability credentials. But I hear bad things about the service it provides. What you hear from children and school staff is that the portions are too small and the children are hungry.
>
> (Interview, 25 June 2021)[3]

The Welsh government has a school food quality standard and is planning to update it. There is no external inspection of the quality of school food. Historically there has been little take up of the Food for Life approach in Wales. As regards levels of funding scarcity of money for high-quality food is as much an issue in Wales as it is in Scotland.

Conclusions: Consistent policies in Scotland and Wales, a political football in England

In all three countries there are catering professionals, NGOs, campaigners, parents, children, councils and politicians who have pushed for better quality school meals, more healthy and sustainable food.

The practical methodology for improving the quality of school meals has shown striking similarities between the UK and other countries such as Sweden or Denmark – cooking from fresh ingredients (scratch) local food sourcing, seasonal menus, reductions in food waste and meat usage.

Government policy in England has been characterised by repeated drastic changes in direction. There have been two phases when government policy supported improving school meals and provided large-scale financial resources and established new standards mandating quality school food – under the Labour government between 2002 and 2010 and under the Coalition, 2013–2015. Between 1980 and 1997 the government's focus overwhelmingly was on keeping down spending on public services – including school meals – and on minimising state intervention in child nutrition. This was also the case with the Coalition, 2010–2013 and to a very large extent with Conservative governments after 2015.

In Wales and Scotland by contrast the devolved governments have shown consistent aspirations to improve school meals. A key issue has been lack of money, with the national government cutting overall funding available for public services.

Notes

1 This table is based on information extracted from the Soil Association website on the four specified dates. The UK has no official statistics relating to organic food in public catering. The only publicly available information has been that released by the Soil Association in relation to the Food for Life Served Here scheme – where Silver and Gold levels require 5% and 15% organic respectively. Table 2.2 was collated by the authors adding up the data about 77 individual local authorities given in print outs of the relevant tables which were made on specific dates.

(www.sacert.org/catering/schoolcaterers

Print outs dated 18 January 2013, 7 February 2015 and 25 February 2016;

www.soilassociation.org/certification/catering/sectors/schools/schools-award-holders/ Print out dated 30 January 2018.)

The 25 February 2016 total of 6,273 schools relates to local authority catering services. The full total of schools with FFL was over 8,000 including private sector catering contractors. It was decided not to include these in this table because the information often cannot be localized to specific local authorities

2 Unpublished interview, 25 March 2021, carried out as part of the research for Mark Stein's PhD thesis. *Sustainable Food Procurement in Public Catering – comparison of the UK with Denmark & Sweden* (2022).

3 Interview for MS PhD research.

References

APSE. (2020, February). *Selling Services to Schools: Facing the Future'*. Retrieved from www.apse.org.uk/apse/index.cfm/news/articles/2020/selling-services-to-schools/#

Belfield, C., & Sibieta, L. (2015). *English Schools will Feel the Pinch Over the Next Five Years* [Comment]. Institute for Fiscal Studies. Retrieved 31 July, 2023, from https://ifs.org.uk/articles/english-schools-will-feel-pinch-over-next-five-years

Bowden, C., Holmes, M., & Mackenzie, H. (2006, 11 August). *Evaluation of a Pilot Scheme to Encourage Local Suppliers to Supply Food to Schools*. Retrieved from https://webarchive.nrscotland.gov.uk/3000/www.gov.scot/Publications/2006/07/27135746/0

Bradshaw, B. (2017, 6 June). *Infant School Meals – A Brief History and Why We Need to Save Them*. Retrieved from https://foodactive.org.uk/infant-school-meals-a-brief-history-and-why-we-need-to-save-them/

Bull, T. (2019, 22 February). *Think of Our Future, GS Plus Staff Plead as Greenwich Council Spin-off Firm Hits Buffers*. Retrieved from https://853.london/2019/02/22/think-of-our-future-gs-plus-staff-plead-as-greenwich-council-spin-off-firm-hits-buffers/

Children's Society. (2012, 13 June). *Briefing. Nutritional Standards for School Food in Academies and Free Schools House of Lords*. Retrieved from www.childrenssociety.org.uk/sites/default/files/tcs/u32/briefings_for_fs_nutritional_standards_debate.pdf

Chiu, B. (2014, 3 July). *Nick Clegg's Free School Meals: A Policy Served Cold*. Retrieved from https://whitehousecomms.com/blog/2014/07/03/nick-cleggs-free-school-meals-a-policy-served-cold/

Compton, R. (2014). *Good Food for London 2014*. Retrieved 20 November, 2020, from www.sustainweb.org/secure/GoodFoodForLondon2014.pdf

Dalmeny, K., Reynolds, B., & Williams, H. (2011). *Good Food for London 2011 October/November 2011*. Retrieved 20 November, 2019, from www.sustainweb.org/secure/GoodFoodForLondon_lowres_nov11.pdf

Davenport, S. (2019, October). *Good Food for London, 2019*. Retrieved 20 November, 2019, from www.sustainweb.org/publications/good_food_for_london_2019/

DEFRA. (Department for Food, Education & Rural Affairs). (2002). *English Organic Action Plan*. Retrieved from https://web.archive.org/web/20080727155906/www.defra.gov.uk/farm/organic/policy/actionplan/pdf/actionplan.pdf

Department for Education. (2010, 15 July). *FOI Release: The Reaction of the British Government to Jamie Oliver's TV Programme on School Meals*. Retrieved from www.gov.uk/government/publications

Dimbleby, H. (2021). *National Food Strategy – Independent Review – Chapter 16*. Retrieved 11 August, 2021, from www.nationalfoodstrategy.org/wp-content/uploads/2021/07/National-Food-Strategy-Chapter-16.pdf

Dimbleby, H., & Vincent, J. (2013, July). *The School Food Plan*. Retrieved from www.schoolfoodplan.com/wp-content/uploads/2013/07/School-Food-Plan-2013.pdf

FFLP. (2010). *Return of the Turkey Twizzler? How Cost Cuts Threaten the School Food Revolution*. Bristol: Soil Association. Retrieved from www.foodforlife.org.uk/LinkClick.aspx?fileticket=GXgaF29JZ78%3d&tabid=118

FFLP. (2011). *Good Food for All: The Impact of the Food for Life Partnership*. Retrieved from www.foodforlife.org.uk/LinkClick.aspx?fileticket=YyUBCvfUWCc%3D&tabid=310

FFLP. (2013). *Food for Life Partnership: How to achieve the Bronze Food for Life Partnership Award Criteria and Guidance*. Retrieved from www.foodforlife.org.uk/~/media/files/general/bronzecga416pp21aug2013web.pdf

FFLP. (2021). *Our Impact Food for Life Served Here, 2020–2021*. Retrieved from www.foodforlife.org.uk/~/media/files/evaluation%20reports/fflsh-impact-report-2021.pdf

Food Foundation. (2022, Autumn). *The Superpowers of Free School Meals Evidence Pack*. Retrieved from https://foodfoundation.org.uk/sites/default/files/2022-11/FSM%20Evidence%20Pack_0.pdf

Food Standards Agency. (2022). *School Food Standards Compliance Pilot Underway In 18 Local Authorities Across England*. Retrieved from www.food.gov.uk/news-alerts/news/school-food-standards-compliance-pilot-underway-in-18-local-authorities-across-england

Guerlain, M. (2018, 5 November). *Good Food for London 2018*. Retrieved 20 July, 2019, from www.sustainweb.org/publications/good_food_for_london_2018/

House of Commons Environment, Food and Rural Affairs Committee. (2021, 3 March). *Public Sector Procurement of Food Sixth Report of Session 2019–21 Report*. Retrieved from https://committees.parliament.uk/publications/5509/documents/54917/default/

Lancashire County Council. (2005, 12 July). *Education Overview & Scrutiny Sub-Committee*. Item 6 Report on Quality of Food in Schools and Child Obesity. (Not available online).

Lancashire County Council. (2022, 27 May). *Local Approach the Order of the Day for School Lunches in Lancashire*. Retrieved from https://news.lancashire.gov.uk/news/local-approach-the-order-of-the-day-for-school-lunches-in-lancashire

Leeds City Council. (2017, November). *Appendix 1 Civic Enterprise Leeds Q2 Reporting*. Retrieved from https://democracy.leeds.gov.uk/documents/s167839/2%20Appendix%201%20-%20Civic%20Enterprise%20Leeds.pdf

London. (2023, 20 February). *Mayor Announces Every London Primary Schoolchild to Receive Free School Meals*. Retrieved from www.london.gov.uk/mayor-announces-every-london-primary-schoolchild-receive-free-school-meals

Manchester City Council. (2020, 1 December). *Resources and Governance Scrutiny Committee – Withdrawal from School Catering Provider Market*. Retrieved from https://democracy.manchester.gov.uk/documents/s21663/Withdrawal%20from%20school%20catering%20provider%20market.pdf

Mikkelsen, B. E., & Sylvest, J. (2012). Organic Foods on the Public Plate: Technical Challenge or Organizational Change? *Journal of Foodservice Business Research*, 15, 1, 64–83. doi: 10.1080/15378020.2011.650541

Morgan, K. (2008). Greening the Realm: Sustainable Food Chains and the Public Plate. *Regional Studies*, 42, 9, 1237–1250. doi: 10.1080/00343400802195154

Morgan, K. (2011, July). Let Them Eat Pickles: How Local Authority Cuts Could Kill the School Dinner Revolution. *Education Journal*, 128, 16.

Morgan, K., & Sonnino, R. (2008). *The School Food Revolution: Public Food and the Challenge of Sustainable Development*. London: Earthscan.

Mottershead, D., & Maréchal, A. (2017). *Promotion of Agroecological Approaches: Lessons from Other Countries*. A Report for the Land Use Policy Group by the Institute for European Environmental Policy (IEEP). Retrieved from http://publications.naturalengland.org.uk/publication/4851581125656576

Nelson, M. (2012, May). *Academies and Compliance With School Food Standards: Research Overview*. Retrieved from www.childrensfoodtrust.org.uk/assets/research-reports/Food_in_Academies_Research_Summary_v2.pdf

Oliver-Larkin, M., & Luck, A. (2020). *Response, Resilience and Recovery: London's Food Response to Covid-19*. Retrieved from https://www.sustainweb.org/publications/response-resilience-recovery-2020/

Parente, S. (2016). *Good Food for London 2016*. Retrieved from https://www.sustainweb.org/secure/GoodFoodForLondon2016.pdf

Pearce, H. (2006). *Setting the Standard. How Food for Life, a Soil Association Pilot Project, Set the Standard for School Meals and Food Education*. Retrieved from www.soilassociation.org/media/4918/policy_report_2006_setting_standard.pdf

Poulter, S. (2012). Michael Gove Under Fire After Allowing Junk Food Back into Schools... While Accepting £50,000 Donations from Dominos Pizza. *Daily Mail*. Retrieved from www.dailymail.co.uk

Price, K. (2023, 29 June). *Uplift to the Universal Infant Free School Meals (UIFSM) Rate for 2023–24 Academic Year*. Retrieved from www.egfl.org.uk/news/2023/06/uplift-universal-infant-free-school-meals-uifsm-rate-2023-24-academic-year#:~:text=An%20uplift%20to%20the%20per,2023%20to%202024%20academic%20year

School Food Matters. (2014). *Evaluation Report School Food Matters' Primary School Meals Improvement Campaign in Richmond – 2007 to 2011*. Retrieved from www.foodmatters.org/wp-content/uploads/2021/07/School-Food-Matters.pdf

Scottish Government. (2002). *Hungry For Success. A Whole School Approach to School Meals in Scotland Final Report of the Expert Panel on School Meals*. Retrieved from www.scotland.gov.uk/Resource/Doc/47032/0023961.pdf

Scottish Government. (2016, 27 January). *Organic Ambitions: Scotland's Organic Action Plan 2016–2020*. Retrieved from https://www.gov.scot/publications/organic-ambitions-scotlands-organic-action-plan-2016-2020/

Soil Association. (2018). *Schools Award Holders*. Retrieved 30 January, 2018, from www.soilassociation.org/certification/catering/sectors/schools/schools-award-holders/

Soil Association. (2019a). *Celebrating 10 Years – Local Authority: Havering*. Retrieved from https://foodforlife.org.uk/~/media/files/fflsh%20and%20gks/ealy%20adopters/havering/havering-case-study_final.pdf

Soil Association. (2019b). *State of the Nation. Children's Food in England, 2019*. Retrieved 1 December, 2020, from www.foodforlife.org.uk/~/media/files/sotn/sotn_2019.pdf

Soil Association. (2019c). *Food for Life Served Here: Handbook 2019- Schools*. Retrieved from www.soilassociation.org/media/20184/fflsh-standards-handbook-schools.pdf

Soil Association. (2021). *Organic Market 2021*. Retrieved 15 October, 2021, from www.soilassociation.org/certification/market-research-and-data/the-organic-market-report-2021/

Sørensen, N. N., Tetens, I., Lassen, A. D., & Løje, H. (2016). *Organic Food Conversion in Danish Public Kitchens: The Effects of the Danish Organic Action Plan 2020 on Organic Public Procurement and Wellbeing at Work*. Søborg: National Food Institute, Technical University of Denmark. Published PhD thesis. Retrieved from http://orbit.dtu.dk/files/124231485/PhD_Thesis_Nina_N_rgaard_S_rensen_print_final.pdf

Tower Hamlets Council. (2019, 27 February). *Report of: Debbie Jones, Corporate Director Children's Services. Recommendations for the Future Delivery of Contract Services – Resolving the Deficit Position*. Retrieved from https://democracy.towerhamlets.gov.uk/documents/s143598/6.5%20Contracts%20Services%20Review%20Report.pdf

Weale, S. (2022, 14 June). *Infant Free School Meals Funding Rises by just 7p a Meal in England*. Retrieved from www.theguardian.com/uk-news/2022/jun/14/infant-free-school-meals-funding-in-england-rises-by-just-7p-per-pupil

2 Sustainable public food procurement in English higher education

Penelope Giosa

Introduction

This chapter focuses on public food procurement in English higher education. It discusses the results of a survey, which involved 19 online interviews with university procurement officers involved in procurement of food and catering services. The research examined the state of sustainable public food procurement practice. It considered challenges and opportunities for embedding sustainability in the procurement process. It discusses practical examples and best practice

The substantial size of the public procurement market and its important role in setting trends that can encourage or discourage sustainable practices justifies the study of sustainable public procurement (Day, 2005). HEIs are significant consumers of resources, generating considerable waste and other environmental impacts (Young et al., 2016). Public catering can play an important role as a driver of sustainable development because it is closely linked with everyday life food consumption and can set a good example for consumers and citizens and thereby affect the supply and demand for sustainable solutions (Wahlen et al., 2012). In the UK, almost one-third of the meals served outside the home daily are by caterers working within the public sector (DEFRA, 2008). It is estimated that the public sector serves 1.9 billion meals a year, which is over 5% of the total UK food service turnover (National Food Strategy, 2021), thus spending annually about £2 billion on public food and catering (House of Commons, 2007; Cabinet Office-The Strategy Unit, 2008).

This survey provides original empirical evidence of how procurement officers in English universities carried out sustainable procurement for food and catering services. It is important to explore the active role of practitioners in reforming existing systems of provision (Shove & Walker, 2010; Wahlen et al., 2012). This chapter highlights the challenges practitioners face when attempting to deliver sustainability "on the ground".

Methodology

This study is based on 19 online interviews with university procurement officers involved in procurement of food and catering services.

DOI: 10.4324/9781003393023-4

The Uni Guide (www.theuniguide.co.uk/) provided a list of 125 English universities. Due to the size of the target population the random number method was used to choose which universities to contact. Each English university was assigned a number. The random number function (RAND) in Microsoft Excel was then used to generate random numbers to select a subset of the target population. It has been suggested that for most research projects aiming to comprehend perceptions and experiences among a group of relatively homogenous individuals, 17 interviews should suffice (Francis et al., 2010). Thirty English universities were randomly chosen and contact made with their procurement officers via their email addresses, which were publicly available on the university websites. Nineteen universities agreed to an interview. The universities participating were the following: Birmingham, Bolton; De Montfort; Lancaster; Lincoln, London Metropolitan, Manchester Metropolitan; Newcastle; Oxford Brookes University, Plymouth; Portsmouth; Reading; Salford; Sheffield; Staffordshire; University College London (UCL); Westminster; York; York St John.

Interviews were conducted online via zoom. Interviews were semi-structured, in an attempt to ensure an interactive conversation that would give the interviewees the opportunity to raise other points which were outside the scope of the interview questions.[1]

There were 13 interview questions in three groups. The questions were based on the Government Buying Standards for Food and Catering Services (GBSF) Checklist, published in 2017 by Public Health England, an executive agency of the Department of Health (Public Health England, 2017).

The first group of questions focused on the existence of minimum mandatory buying standards in tender specifications and contract performance conditions for food and catering services. The questions in this group included both yes/no queries as well as a request for specific examples. The questions on the minimum mandatory buying standards related to the following areas: a) food production, processing and distribution, b) animal welfare and c) environment. Regarding food production, processing and distribution interviewees were asked "are there any minimum mandatory buying standards for food and catering, which are relevant to the traceability of fresh, chilled and frozen produce?" If so, "please, briefly explain what exactly these minimum mandatory buying standards include". Regarding animal welfare interviewees were asked "are there any minimum mandatory buying standards for food and catering, which are relevant to animal welfare?" If so, "please, briefly explain what exactly these minimum mandatory buying standards include". Regarding environment, the question was "are there any minimum mandatory buying standards for food and catering, which are relevant to the reduction of environmental impacts or high environmental production standards?" If so, "please, briefly explain what exactly these minimum mandatory buying standards include".

The second set of questions focused on the best practice standards that university procurement officers had in respect procurement of food and catering services. The question asked at this point was open-ended in an attempt to understand the factors/parameters relevant to sustainability that they usually considered before reaching an agreement with suppliers in procurement of food and catering services.

The third major topic was about the university's ongoing procurement agreements for food and catering services. It first centred on the theoretical possibility of using sustainability criteria as part of the procurement specification and selection of their current food and/or catering services suppliers: "Have you used sustainability criteria as part of the procurement specification and selection of your current food and/or catering services suppliers? If so, what sustainability criteria have you set?" Then, the questions focused on specific issues, such as the commitment of current suppliers to waste and emission reduction policies and standards, the inspection of current and past suppliers regarding the use of reusable/recyclable packaging for their products, the methods of recording and monitoring food waste in the campus cafes, the policy for the disposal of excess, unserved food, as well as the procurement of organic food.

The survey results and discussion

Minimum mandatory buying standards

Several scholars have alleged that procurement managers and policy makers have largely failed to factor sustainability criteria into the public purchasing equation (Sonnino & McWilliam, 2011; Morgan, 2008; Lang, 2010). Interviewees were asked if any sustainability criteria had been used within the procurement specification and selection of the current food and/or catering services suppliers.

The survey showed that in 5 out of 19 universities (26.3%) had not used sustainability criteria in their tender documents. Instead they had asked open questions regarding various sustainability issues and then considered the relevant answers. For example, Salford asked potential suppliers to outline the steps that the tendering company had taken towards reducing its environmental impact and how it would seek to minimise the environmental impact of services to be provided within the contract.

The survey showed that 14 out of 19 universities (73.7%) had included sustainability considerations in the tender specifications. The inclusion of these criteria in contractual arrangements with the suppliers will have encouraged the latter to transition into more sustainable practices. This may be seen as a positive development. A new philosophy was being promoted in university food procurement, according to which every participant, from the contracting authority to consumers, understood and took responsibility for the environmental consequences of food production, services and consumption.

Along the same lines, it was further asked whether sustainability had any minimum percentage (%) in the tender documents of each university. Those universities that did not set any sustainability criteria in their tender documents did set a minimum percentage (%) weighting for sustainability. Most of the universities that ensure sustainability had a specific percentage (%) weighting in their tender documents set quite a high threshold, that is 10% and above. Only three universities had a sustainability percentage weighting below 10%. The highest threshold stated by procurement officers was 25%.

The vast majority of the universities which took part in the survey bought at least some of their food requirements through university purchasing consortia. University purchasing consortia are buying organisations, which develop and manage framework agreements so that subscribing universities have access to professionally tendered agreements (University of Bath, 2019). By using those framework agreements, universities can purchase competitively priced goods and services across a wide range of commodity areas. This also helps them to carry out all the necessary checks in order to ensure that suppliers have a sustainable food policy and comply with food safety standards.

Particularly, when it comes to food production, processing and distribution, the purchase of food and catering services, which are endorsed by a recognised certification scheme, such as Red Tractor, LEAF Marque, Marine Stewardship Council, Lion Mark, Food for Life Catering Mark and Fairtrade mark, helped procurement officers in the tendering process to assess the commitment of bidders to sustainable food policy and their compliance with the specifications set in the contract notice. For this reason, but also because some of the aforementioned logos promote animal welfare, there is great reliance on and strong preference for the aforementioned certification schemes by the universities that took part in the survey. Fairtrade food is particularly popular among English universities, as 16 out of 19 surveyed (84.2%) had fair trade products on campus, such as coffee, tea, chocolate and bananas. Nevertheless, some interviewees acknowledged that the aforementioned standards have their limitations. Therefore, they avoided specifying in their contract notices the aforementioned certification schemes so as not to stifle other similarly good initiatives and particularly those which might promote short supply chains.

In case of food, which was not endorsed by a recognised certification scheme, Newcastle University asked its suppliers to provide a full and structured risk assessment that demonstrated the food was sourced from a management regime that met the minimum requirements for the product group. This was a good practice that could be taken forward by other English universities. Other good practice regarding the production standards of the food purchased were the monthly reports that UCL asked its suppliers to give about the origin of their meat, the annual demonstrations of sourcing (particularly sustainable and ethical fish) that UCL required on an audit basis and the Flexible Framework, a self-assessment mechanism developed by the business-led Sustainable Procurement Task Force, which the University of Birmingham used in order to measure and monitor its progress on sustainable procurement over time. Sheffield had partnered with a sustainable coffee sourcing business, which bought coffee from a growers' cooperative in Colombia. Sheffield were thus able to identify where their coffee beans are grown and processed. Roasting took place only three miles from campus.

The minimum mandatory buying standards relevant to the reduction of environmental impact mainly regarded plastics usage and food miles as well as compliance with ISO 14001 which required suppliers to reduce environmental impacts through an environmental management system. Birmingham set up a mandatory buying standard similar to compliance with ISO 14001. It required tenderers to adopt the Net Positive Futures approach, according to which apart from reducing negative environmental impacts, businesses should also enhance the positive impacts of their activity – environmental, social and economic.

In their attempt to reduce plastics usage, many participating universities used wooden cutlery, paper bags, cardboard salad containers, cans instead of plastic bottles and promoted the use of own coffee mugs incentivised with discounts. Indicatively, Sheffield Students' Union introduced a coffee cup levy for single-use coffee cups with funds ring-fenced to fund Students' Union sustainability projects. UCL and Oxford Brookes asked their suppliers to commit to reduce single-use plastics. UCL aimed to have a plastic free campus by 2024. Portsmouth asked their fruit suppliers to use cardboard or reusable instead of single-use plastic. York St. John used key performance indicators (KPIs) to reduce usage of plastics. However several interviewees acknowledged that due to COVID-19, the university had reverted to single-use plastics, but they were monitoring their performance and trying to reduce usage once again.

With regard to food miles, almost all the universities cooperated with local and small suppliers. An illustrative example was Reading, which explored cooperation opportunities with the Federation of Small Businesses. The project involved the identification of local and small businesses and the development of a tendering process, which was fair and transparent and relatively easy to complete. Alongside this, Reading explored opportunities to create a supplier development programme, which would help small business to become more sustainable. Newcastle and London Metropolitan also set an example. Newcastle required as a minimum standard that the suppliers agree a delivery schedule with the university to minimise the frequency of deliveries to the catering units. London Metropolitan included in the tender process requirements for food traceability, such as the percentage of local food and how far from the university the food was produced. Instead of minimum buying standards, there were also universities that set a policy objective to increase the proportion of expenditure on local suppliers and they considered the locality as a parameter in the tendering process, without being bound by it, as they would be if it were a minimum buying standard specified by the contracting authority. For instance, Oxford Brookes aims at suppliers within a 60-mile radius. Likewise, Plymouth tries to collaborate with local companies even if they are more expensive. Indicatively, 48% of their products are from Devon and Cornwall.

Several interviewees raised concerns regarding the impact of collaborative procurement on the inclusion of local, small and medium enterprises. The framework agreements established by university purchasing consortia provide food at attractive prices and good safety standards. However, they are a big obstacle with respect to buying from local and small suppliers. SMEs have difficulties engaging with framework agreements because they are usually very large contracts, the bidding requirements are disproportionate and SMEs lack the resources to service these complex and burdensome procurement procedures.

Best practice standards

A systematic literature review of sustainable public food procurement of food, recommended that further research should gather more examples of best practice (Molin et al., 2021). Interviewees were asked about their best practice standards regarding procurement of food and catering services. The majority of interviewees answered that beyond the minimum mandatory buying standards in the aforementioned areas, they also included requirements related to the management of slavery risks. Indicatively, Reading required that potential suppliers disclose information on prior violations and ensure transparency in their supply chain. UCL required that its suppliers describe the steps – if any – they had taken to eradicate slavery from their supply chain. Reading had extensive audit arrangements to identify any illegal practices. In UK law, modern slavery is an umbrella term, which encompasses the offences of slavery, servitude, forced or compulsory labour and human trafficking (MSA, 2015). The interviewees' determination to tackle modern slavery crimes complies with the assertion that public procurement is a "mechanism by which governments can leverage corporate responsibility for labour standards, and can be used as a method either to enforce existing legal obligations or advance desired modes above and beyond those legally required by law" (Emberson & Trautrims, 2020; See also Howe, 2010). Interviewees also mentioned that they asked their suppliers to provide evidence of their commitment to the creation of apprenticeships and student employment opportunities under the contracts.

A good example is York St John University, where in the context of a high-profile project called "The Living Lab" over 800 students had the opportunity to work collaboratively with academic staff and professional teams in order to help the university get the Soil Association Food for Life status. To this aim, indicatively students were engaged with food growing on campus, they investigated the ethics of the catering operation's suppliers, they endeavoured to develop more easily recyclable pizza boxes and promoted plant-based options and they edited the York St John Student Cookbook, which included not only recipes but also stories of students' food memories, identities, values and challenges.

Ongoing procurement agreements for food and catering services

The last part of the interviews was dedicated to the universities' ongoing procurement agreements for food and catering services. It was firstly asked if the commitment of their current suppliers to waste and emission reduction policies and standards has been checked. The survey shows that the majority of English universities had checked the commitment of their current suppliers to waste and emission reduction policies and standards (16 out of 19 universities, 84.2%). This has been done in supplier review meetings, via a vetting procedure, which included scrutiny of the suppliers' sustainability practices or by companies that check this on behalf of the universities. Staffordshire also mentioned the Net positive supplier engagement software, which is a tool that has offered their suppliers an opportunity to generate their own Sustainability Action Plan (https://netpositivefutures.co.uk/netpositive-supplier-engagement/). Based on data generated by their suppliers,

Staffordshire assesses how suppliers are progressing in waste and emission reductions and is also more able to understand their needs.

When discussing waste reduction policies and standards, most interviewees referred to the use of reusable/recyclable packaging for purchases and to recycling policies. Almost all participating universities (17 out of 19, 89.5%) assess their suppliers regarding the use of reusable/recyclable packaging by conducting on-site inspections/physical checks, by reviewing supplier action plans or by discussing with suppliers in review meetings. An illustrative example is York St. John University, which changed its waste management company because they could not recycle a wide enough range of materials. Another illustrative example is University of Westminster, which has set the Green Kitchen standard as an incentive for its suppliers to minimise environmental impact by reducing energy and water usage. Green Kitchen Standard is an accreditation from Food for Life, which can be awarded to sustainable catering operations which have taken steps to reduce food waste and energy and water usage (Food for Life, 2019.). York marked the bidders during the procurement procedure on their ability to reduce packaging waste, while Newcastle required that suppliers take back all packaging within an agreed period. In the same vein, instead of using plastic containers for the delivery of milk, Sheffield set up a delivery system for milk, which recycled stainless steel containers.

Regarding emission reduction and energy management policies, only 5 out of 19 (26.3%) of the universities assessed suppliers' policies regarding energy efficiency as part of the tender evaluation. Several interviewees explained that universities' current influence over suppliers is rather limited, so they cannot do much to modify the energy consumption and resource efficiency of their suppliers. Most of the universities participating in the survey consider energy efficiency as part of the tender evaluation, only when it comes to the purchase of equipment. In that case, universities' procurement officers assess the energy and utility consumption of the equipment they want to buy. Specific reference was made to the purchase of the most energy-efficient cooking or catering equipment and vending machines.

In terms of food waste, only one university in this survey made it clear that the caterers' monitoring system was a requirement in the tendering process. As regarded disposal of unserved/excess food, the survey showed that participating universities did their best to minimise food waste through involvement in numerous initiatives. These included the operation of a community fridge in the residences so that anyone could get food for free and collaboration with the mobile application Too Good To Go, which was a service with a mobile application that connected users with businesses that had unsold surplus food at the end of the day. The universities also donated unserved food to food banks, hostels or charities. Several composted their food waste to use as fertiliser or turned it into biofuel to produce renewable energy. For instance, Portsmouth composts food waste whereas De Montfort transfers its food waste to local company for use as fuel. At Reading and Oxford Brookes food waste is segregated and collected for anaerobic digestion.

Eight universities monitor demand in their dining facilities in order to reduce excessive food preparation and ingredient purchases. Portsmouth collaborates with a company, which measures food waste and suggests ways to reduce it. At De

Montfort kitchen waste is weighed and students are able to bring their own food waste to recycle it with the university's waste. London Metropolitan University has monthly meetings with catering manager to assess the weight of waste produced and the sources of waste so that they can set smart goals to reduce waste.

The higher cost of organic food, especially organic meat, inhibited purchases because of the need to keep down the price of meals charged to students. Six universities were in favour of purchasing organic food. Reading was moving to organic pork and working on establishing a local organic chicken supply. UCL, Salford and Reading bought organic milk and vegetables. Westminster, London Metropolitan and Manchester Metropolitan aspired to include more organic fresh produce in their menus. Several interviewees said that reduction of meat usage was a crucial way to finance more sustainable food, including organic.

Conclusions and recommendations

The first conclusion of this research is that English universities support sustainability practices and they demonstrated their commitment to them by

a) buying food and catering services endorsed by recognised certification schemes,
b) procuring local food and food from SMEs,
c) reducing plastic usage and recycling plastics,
d) managing slavery risks,
e) supporting skills development and apprenticeships, and
f) requiring efficient use of resources and reduction of waste from their suppliers.

The fact that 14 out of 19 universities (73.7%) interviewed had minimum mandatory buying standards relevant to environmental, social and economic sustainability showed the significance of setting forth such requirements in the tender documents and contracts with suppliers. In this way the public procurement officers have steered their suppliers towards environmental, social and economic sustainability.

Many interviewees however felt that their bargaining power as an individual university was limited compared to that of the larger food suppliers. For this reason, only 5 out of 19 (26.3%) universities felt able to ask their potential suppliers as part of the tender evaluation to demonstrate their ability to reduce emissions and manage energy efficiently. This issue could be addressed by encouraging more universities to buy food and catering services through university purchasing consortia. This could increase their bargaining power and would also give them access to highly experienced procurement and supply chain specialists employed by these consortia.

The second conclusion of the research was that food and catering services endorsed by a recognised certification scheme were regarded by most interviewees as a trustworthy assurance of quality and sustainability. However, some interviewees acknowledged the limitations of these certifications. They may encourage certain standards to be upheld, but their use is not a guarantee that will always prevent the greatest risks in their supply chains, such as human rights violations and environmental harm. Procurement officers could pay greater attention to risk assessments carried out by these certifying bodies to mitigate these risks.

The third conclusion relates to food miles and the use of public food procurement to support small producers and SMEs. Since several interviewees raised concerns regarding the impact that framework agreements made by university purchasing consortia may have on the inclusion of local, small and medium enterprises, it is appropriate to discuss ways to deal with this challenge. Therefore, it is suggested that English universities should consider the breakdown of their procurement contracts into smaller lots in order to allow smaller suppliers to bid for certain parts of their supply contracts. This could be done by dividing contracts either across geographic lots or into product-specific lots. The size of lots should be heterogeneous and their value should be asymmetric in order to lower the barriers to market entry and open up the market to SMEs, without favouring any attempt of market sharing between the bidders (OECD, 2018). However, procurement officers should always bear in mind that the aforementioned rule cannot apply effectively to anti-competitive practices, such as cover bidding, bid rotation and subcontracting arrangements, so they should be cautious about potential infringements of competition law by their bidders, when they divide contracts into lots (Giosa, 2018).

A last point that is worth discussing is how English universities could possibly overcome difficulties in providing organic food on their campuses because of its high cost. A possible way is to rethink menus, with less frequent usage of expensive ingredients – particularly meat and fish. This could compensate for the higher price of organic food products and could keep the costs within budget, according to Risku-Norja and Løes (2017). In the same vein, Nuutila et al. (2019) conclude that "menu engineering could provide a solution for public catering that faces the challenge to procure and offer food services more sustainably". Another suggestion is the seasonal use of organic food, as this could enable contracting authorities, such as English universities to buy organic food directly from producers without intermediaries. A short organic supply chain may be able to offer food at a price competitive with long conventional supply chains (Carmona et al., 2011).

Note

1 All interviews were carried out using Zoom except for one which was conducted via email, as the relevant procurement officer did not have enough time for an online interview but was able to respond in writing to the research questions. Email interviewing is another reliable method of data collection (Hawkins, 2018; James, 2016; Wertz et al., 2011).

References

Cabinet Office-The Strategy Unit. (2008). *Food matters: towards a strategy for the 21st century*. Retrieved from https://webarchive.nationalarchives.gov.uk/+/http:/www.cabinetoffice.gov.uk/media/cabinetoffice/strategy/assets/food/food_matters1.pdf

Carmona, I., Griffith, D. M., & Aguirre, I. (2011). 'Understanding the factors limiting organic consumption: the effect of marketing channel on produce price, availability, and price fairness', *Organic Agriculture* 11, 89–103.

Day, C. (2005). 'Buying green: the crucial role of public authorities', *Local Environment* 2(10), 201–209.

Department for Environment, Food and Rural Affairs (DEFRA). (2008). *Food statistics pocketbook*. Retrieved from https://webarchive.nationalarchives.gov.uk/ukgwa/20130103024812/www.defra.gov.uk/statistics/foodfarm/food/pocketstats/

Emberson, C., & Trautrims, A. (2020). 'How might modern slavery risk in English adult social care procurement be reduced?' *Public Procurement Law Review* 6(29), 380–394.

Food for Life. (2019). *Green kitchen standard*. Retrieved 30 April, 2023, from www.foodforlife.org.uk/catering/green-kitchen-standard

Francis, J. J., Johnston, M., Robertson, C., Glidewell, L., Entwistle, V., Eccles, M. P., & Grimshaw, J. M. (2010). 'What is an adequate sample size? Operationalising data saturation for theory-based interview studies', *Psychology and Health* 25, 1229–1245.

Giosa, P. (2018). 'Division of public contracts into lots and bid rigging: can economic theory provide an answer?', *European Procurement and Public Private Partnership Law Review* 13(1), 30–38.

Hawkins, J. E. (2018). 'The practical utility and suitability of email interviews in qualitative research', *The Qualitative Report* 23(2), 493–501.

House of Commons. (2007). *Smarter food procurement in the public sector: thirteenth report of session 2006–07*. Retrieved from https://publications.parliament.uk/pa/cm200607/cmselect/cmpubacc/357/357.pdf

Howe, J. (2010). The regulatory impact of using public procurement to promote better labour standards in corporate supply chains. In Macdonald, K., & Marshall, S. (eds), *Fair trade corporate accountability and beyond: experiments in global justice* (1st ed.), 329–347. Ashgate.

James, N. (2016). 'Using email interviews in qualitative educational research: creating space to think and time to talk', *International Journal of Qualitative Studies in Education* 29(2), 150–163.

Lang, T. (2010). 'From 'value-for-money' to 'values-for-money'?' Ethical food and policy in Europe. *Environment and Planning* 42(8), 1814–1832.

Modern Slavery Act. (2015). *An act to make provision about slavery, servitude and forced or compulsory labour and about human trafficking, including provision for the protection of victims; to make provision for an independent anti-slavery commissioner; and for connected purposes*. Retrieved from https://www.gov.uk/government/collections/modern-slavery-bill

Molin, E., Martin, M., & Björklund, A. (2021). 'Addressing sustainability within public procurement of food: a systematic literature review', *Sustainability* 13, Article 13395. doi: 10.3390/su132313395

Morgan, K. (2008). 'Greening the realm: sustainable food chains and the public plate', *Regional Studies* 42(9), 1237–1250.

National Food Strategy. (2021). Retrieved from www.nationalfoodstrategy.org/wp-content/uploads/2021/07/National-Food-Strategy-Recommendations-in-Full.pdf

Nuutila, J., Risku-Norja, H., & Arolaakso, A. (2019). 'Public kitchen menu substitutions increase organic share and school meal sustainability at equal cost', *Organic Agriculture* 9, 117–126.

OECD. (2018). *Second public procurement review of the Mexican institute of social security (IMSS): reshaping strategies for better healthcare* (1st ed.). OECD Publishing.

Public Health England. (2017). *Government buying standards for food and catering services (GBSF) checklist*. Retrieved from https://assets.publishing.service.gov.uk/government/

uploads/system/uploads/attachment_data/file/595129/Healthier_and_more_suistainable_GBSF_checklist.pdf

Risku-Norja, H., & Løes, A. K. (2017). 'Organic food in food policy and in public catering: lessons learned from Finland', *Organic Agriculture* 7, 111–124. doi: 10.1007/s13165-016-0148-4

Shove, E., & Walker, G. (2010). 'Governing transitions in the sustainability of everyday life', *Research Policy* 39, 471–476.

Sonnino, R., & McWilliam, S. (2011). 'Food waste, catering practices and public procurement: a case study of hospital food systems in Wales', *Food Policy* 36, 823–829.

University of Bath. (2019). *How to do business with the University of Bath.* Retrieved from www.bath.ac.uk/publications/how-to-trade-with-the-university-of-bath/attachments/trading_with_UoB_mar19.pdf

Wahlen, S., Heiskanen, E., & Aalto, K. (2012). 'Endorsing sustainable food consumption: prospects from public catering', *Journal of Consumer Policy* 35, 7–21.

Wertz, F. J., Charmaz, K., McMullen, L. M., Josselson, R., Anderson, R., & McSpadden, E. (2011). *Five ways of doing qualitative analysis* (1st ed.). The Guilford Press.

Young, S., Nagpal, S., & Adams, C. A. (2016). 'Sustainable procurement in Australian and UK universities', *Public Management Review* 18(7), 993–1016.

3 Public procurement and local food in Sweden, Denmark and the UK

Mark Stein and Yiannis Polychronakis

Introduction

This chapter explores public sector approaches to sourcing local food in Sweden, Denmark and the UK. Since the introduction of the Single European Market in the late 1980s EU procurement law has required public authorities to open up procurement opportunities to suppliers from all over the EU (Martin et al., 1999). The UK left the EU on 31 January 2020 but continues to be subject to the World Trade Organisation Agreement on Government Procurement which means that foreign suppliers are still entitled to equal access to all public procurement tenders (Kimbell, 2023). The chapter shows how public sector procurers have found ways of sourcing local food while complying such legal provisions. It reveals significant differences in the approaches developed in the three countries.

The chapter explores different definitions of "local food" which have been used in different times and places. A widespread approach has been to regard "local" as within the boundaries of a local authority area. Another approach – particularly relevant to cities with little or no agricultural land – has been to focus on sourcing from regional suppliers. A third approach has been to examine how much food is sourced within a country's national boundaries.

The chapter will consider the scale of public food procurement arrangements. At one extreme these may be devolved down to the level of individual establishments – such as a school. At the other extreme there may be national food contracts – such as those in Denmark, Scotland and Wales – and a most important question here is the extent to which they prevent any purchasing of food from local suppliers. In between there may be relatively small local authority purchasing arrangements – for up to 200,000 population – which are the typical arrangements found in Sweden. In the UK local authority purchasing arrangements include large-scale regional consortia – which may cover local authorities with over 1.5–2 million population. They also include large individual local authorities. For example, Hampshire, Lancashire, Leicestershire and North Yorkshire with populations ranging between 615,000 and 1.2 million.

The chapter also explores the use of an innovative technique – dynamic food procurement – whereby new and smaller suppliers are able to enter the procurement arrangement much more readily. This is by contrast to the traditional tendering

DOI: 10.4324/9781003393023-5

system which favours larger suppliers and where once the tender has been awarded no new suppliers will be considered for the multi-years duration of the supply contract.

Academic literature

Academic studies of public food procurement have frequently noted that public agencies prefer when possible to source food from local suppliers (Goggins & Rau, 2016). There is very limited academic literature on the techniques used by public agencies to promote local sourcing. Harland et al. (2019, p. 2) observes that engaging "small business in public procurement is critical, difficult and understudied". EU public procurement rules permit buyers to conduct market research and talk informally to potential suppliers prior to issuing the formal tender (McKevitt & Davis, 2015, p. 79). Below a certain threshold public bodies are permitted to carry out procurement without a tender – often referred to as "direct procurement" (Morgan & Sonnino, 2008). Buyer-supplier interaction is not constrained by the EU tender rules. (For more on these rules see European Commission, 2018).

United Kingdom

Public Sector Food Procurement initiative, 2003–2009

The UK was the first of the three countries to launch a national initiative aimed at promoting public procurement of local food. The timing reflected a perceived need to boost agriculture in the wake of the 2001 foot and mouth outbreak.

The Public Sector Food Procurement Initiative (PSPFI) was specific to England – although similar activities did take place in Scotland and Wales. Aims included improving the sustainability of production, processing and distribution, increasing tenders from small and local producers and promotion of organic food and animal welfare (Deloitte, 2009, p. 3).

PSPFI was led by the Department of the Environment, Food and Rural Affairs (DEFRA) and received considerable funding for its activities. Activities included seminars for public sector food procurers, outreach to food suppliers, publication of case studies and progress reports to highlight activities and publication of detailed guidance documents – such as Defra (2006) and DEFRA (2008a).

PSPFI achievements

By October 2008 over a hundred case studies relating to public food procurement by local authorities or other public bodies had been published on the DEFRA website (DEFRA, 2008b). The titles of these case studies reflect a great deal of detailed work which had taken place to promote local food procurement while ensuring high quality standards and working within budget constraints. The case studies referred to numerous successes with sourcing local food for schools and hospitals. Deloitte produced an evaluation of PSPFI in 2009 which concluded that: there is evidence of limited progress towards PSFPI goals: while acknowledging that it was difficult to quantify PSFPI's success because it had not been measured or monitored regularly during its lifetime.

Lessons: how to promote local food sourcing

Michaels (2006) summarised best practice developed by local councils to promote local suppliers. There needed to be outreach to smaller suppliers. Contracts should be split up by geography or product (lotting). Help was needed with logistical arrangements – for example using another, larger contractor to assist with distribution. Best practice in animal welfare should be supported. If there is a switch to buying seasonal fresh produce available locally, menus needed to be changed and staff retrained to cook with fresh ingredients. It was also important to raise awareness of healthy and sustainable food among school children and parents (Michaels, 2006, pp. 1–5).

The new government which came to power in May 2010 initiated massive public expenditure cuts. PSPFI was one of many public sector initiatives which was discontinued to save money.

2010: Proposed national contract

A longstanding policy within UK public procurement has been to seek opportunities to aggregate demand into fewer and larger contracts, which should enable procurers to drive down prices (National Audit Office, 2006, p. 12).

In July 2010 the Department for Education put forward a proposal for a national contract for all public food procurement. Local authorities opposed this because it would dissolve their existing local procurement arrangements which supported local suppliers. Before the end of 2010 the government decided not to proceed with the proposal.

Food for Life, 2007–2018

The Food for Life Partnership [FFLP] was launched early in 2007. This is a voluntary sector initiative promoting sustainable food procurement – particularly in schools but also in other public and private sector settings. It originated as a partnership of four NGOs: Soil Association, Focus on Food, Garden Organic and the Health Education Trust

Food for Life in public catering had many resemblances to PSPFI in encouraging local and in some cases organic food, higher animal welfare, seasonal menus and cooking with fresh ingredients. The FFL approach involved a highly structured scoring system whereby school catering organisation were inspected annually by a visit from the Soil Association inspector who awarded points according to the criteria set out in the FFL Standards Handbook – which for schools was 49 pages long (Soil Association, 2019). The inspectors visited kitchens and dining rooms and examined menus and a sample of purchasing records to verify organic and other accreditations. To encourage purchasing of local and UK food, caterers were awarded 3 points per % of their spend on raw ingredients produced in their region and two points per % of spend over 59% spent on raw ingredients produced in the UK (Soil Association, 2019, p. 37).

The New Economics Foundation carried out a Social Return on Investment study for two areas – Plymouth and Nottinghamshire – which quantified the benefit to local economies from increased spending on local food due to adoption of FFLP practices. In Nottinghamshire the share of ingredient spend on seasonal, local produce in school meals had risen dramatically, by £1.65 million per year, returning £3.11 in social, economic and environmental value for every £1 spent. In Plymouth, the change in spending on seasonal, local produce was valued at £384,000 per year, which was found to generate £1.2 million of value per year (Kersley, 2011, p. 3).

There was a very large increase in Food for Life accreditations between 2011 and 2018. This reflected a change in national government policy in 2014 aimed at improving the quality of school food through the School Food Plan and introduction of Universal Infant Free School Meals. This involved a substantial increase in national government funding for school meals. Another factor which promoted improvements in school meals was the outrage over the fraudulent and illegal usage of horse meat in school meals and other catering during 2013.

Local procurement: Case studies

Several English local authorities have a strong track record in buying food for their school meal services from local and regional suppliers. These include Hampshire, Lancashire, Leicestershire and North Yorkshire (Hampshire County Council, 2019; Lancashire County Council, 2018; Foster, 2018; Sustain, 2021). The large size of these local authorities however means that they contract with large wholesalers to purchase their food. The size of these contracts is such as to exclude small suppliers from bidding. The wholesalers will of course purchase some of their requirements from smaller and local suppliers and add a markup when selling on to their final customers.[1]

The one local authority which has adopted a very different approach was Bath & North East Somerset, which in 2016 decided to set up a dynamic food procurement (DFP) system to buy food for its 70 schools. A contract was signed with a software provider, Equilibrium, to establish an online food store called Fresh Range to carry out order consolidation and delivery. Suitably qualified suppliers – who complied with minimum UK standards – were allowed to join the contract at any time. They could offer what food they had available, which need not be enough to supply all schools. All qualifying suppliers participated in quarterly mini-competitions., Registered suppliers submitted prices for a list of products. Suppliers could also inform buyers about other quality aspects of their food – such as organic food – which was important for FFL certification. Orders made by school kitchens were consolidated by Equilibrium so that each school received only one delivery. This approach was successful in providing the school food service with high-quality, affordable local food. Excluding wholesalers offset higher farmgate prices from local producers. (Soldi, 2018, pp. 23–25; SPP Regions Consortium, 2017; House of Commons, 2020).

British Food Plan, 2014

In 2014 the UK government published a new strategy for public food procurement, the British Food Plan (Bonfield, 2014). This supported a decentralised approach whereby individual local authorities were free to carry out their own procurement strategies. It supported the Food for Life approach and encouraged increased purchasing from small local producers, through such techniques as lotting and supplier engagement. However, the responsible government department, DEFRA, had very few resources to promote more local food procurement. Increasingly short of money, local authorities have paid little attention to the British Food Plan (Soil Association, 2020b).

Deteriorating school food standards, 2018–2023

The deterioration of school food standards from 2018 can be attributed to several factors. School budgets have been reduced. National quality standards have not been enforced. Many schools have withdrawn from local authority catering organisations and brought in cheaper private contractors.

Local authority catering services have shrunk and several which strongly emphasised sourcing of local/regional food have closed down. These included Bath & North East Somerset and Cambridgeshire in 2018 and Manchester in 2021. Those local authority caterers which survive have been under extreme pressure to cut costs (Soil Association, 2020b).

Caterers have increasingly been forced to buy the cheapest food often from distant suppliers, frequently outside the UK. The RSPCA has complained that schoolchildren and hospital patients were eating from “low welfare menus” containing food imported to lower standards than the UK (Farming UK, 2021).

A single school catering organisation buying for all schools in a local authority area was in a much better market position to buy high-quality food – perhaps local and/or organic – than a fragmented set up where school kitchens are run by individual schools or one of a large number private catering contractors.

Scotland and Wales

By contrast to England there has been practically no outsourcing of school catering to private sector in Scotland and very little in Wales. In both countries a national food procurement contract has been set up by the devolved government, which gives local authorities access to low cost and high quality. However, individual local authorities have retained flexibility to procure from local suppliers.

The Scottish government has for several years funded Soil Association Scotland to encourage local authorities to adopt the Food for Life approach for school catering and the number of local authorities with FFL accreditation has now risen to 17 out of 34. The Good Food Nation Act became law in Scotland in June 2022 and one of the objectives is to encourage public procurement of local food.

In Wales there are a number of local initiatives aiming to promote local food procurement. For example the establishment of two Procurement Hubs in Carmarthenshire and North Powys which will work with local, small-scale growers to aggregate supply and provide produce to the public sector that meets its requirements in terms of ease, cost and sustainability (Farmgarden, 2023; Wales Rural Network, 2023).

Dynamic food procurement (DFP), 2019–2023

Over the last four years there have been proposals to transform public food procurement through dynamic food procurement – along the lines of the Fresh Range procurement platform which was provided by Equilibrium software to Bath and North East Somerset Council between 2016 and 2018 and then to some individual schools until October 2020 (see earlier sections).

Enabling local food producers to sell directly to public procurers would bypass the food wholesalers who have long dominated the public sector food market and who have extracted large profits from their intermediary role.

During 2020 and 2021 the idea was taken up by Crown Commercial Services (CCS) which is an executive agency of the Cabinet Office of the UK government. It was proposed to organise firstly a regional pilot within South West England to try out the software platform and see how it worked with the delivery contractor. If the pilot was successful the government would roll the system out nationally (Crown Commercial Services, 2020; 2021).

In March 2021 the House of Commons report on public food procurement expressed support for dynamic food procurement but also concern over doubts as to whether public money was still available for this project after the massive spending incurred by the COVID-19 emergency (House of Commons Environment, Food and Rural Affairs Committee, 2021).

By the end of 2021 it was evident that CCS was not going ahead with its plans. A group of interested individuals and organisations had set up a Dynamic Procurement Steering Group and this had obtained a modest amount of funding from the Dixon Foundation to set up a number of small pilot projects to demonstrate the value of this approach (Dynamic Procurement UK Food, 2022).

Widespread hopes have been expressed that DFP could be a game changer in assisting small and local food producers to sell high-quality foods into public procurement (Soil Association, 2020a; Lawes-Johnson & Woodward, 2022; Morley, 2023; Social Farms & Gardens, 2023).

Aspirations to promote local procurement in England

There continue to be local authorities, public procurers, farmers and NGOs who wish to encourage more public procurement of local food in England. This can be seen in a series of studies published in recent years. Soil Association (2020a) recommended that

- Government should recognise and pay for the 'public goods' delivered by sustainable and regional food production
- National and Local Governments should incentivise sustainable food sourcing and development of short food supply chains in the public sector
- Local government should support the development of regional food supply infrastructure as part of its climate change mitigation strategy (such as local abattoirs, vegetable washing facilities)
- Local food hubs are should be supported to provide IT and logistics capabilities to assist local food producers – including support for dynamic food procurement
- Investment should be made in facilitating the coordination and mapping of local food producers in pilot areas, where public procurement bodies are willing to prioritise procurement of local, sustainably produced foods

Wilkinson et al. (2022) considered the potential for increasing public sector procurement from regional food suppliers in South West England. The study was supported by local authorities and enterprise partnerships in South West England. The report concluded that many smaller, regionally based businesses faced considerable barriers and challenges in engaging with public sector food procurement. A major culture change was needed across the production, processing, wholesale and procurement sectors, with support from both national and local government.

Lawes-Johnson & Woodward (2022) examined investment needed to promote localised food systems in two English counties – East Sussex and Lancashire. The research aimed to assess what infrastructure there was, how it contributed to the local food system, what farm types there were in the counties, and how much investment was needed to fix infrastructure gaps.

Morley (2023) recommended that government should focus on developing fruit and vegetable production in English regions to meet demand from public procurement. There should be additional government support to encourage horticulture producers who will sell into the public sector.

Consequences of Brexit

Entry restrictions on farm labour have made it more difficult for some UK farmers to produce a crop – which has reduced the availability of British produce and made it more difficult for procurers to buy British

In December 2021 the UK government entered into its first wholly new post-Brexit trade agreement. This was with Australia. It was criticised by the UK National Farmers Union because it removed restrictions on Australian food exports to the UK – which would compete directly with food, particularly meat – produced by UK farmers (National Farmers Union, 2023; Mason, 2022). The agreement came into force in May 2023.

June 2022 DEFRA propose 50% local produce target

In June 2022 as part of a consultation on revisions to *Public sector food and catering policy for England The Government Buying Standards for Food and Catering Services (GBSF)* DEFRA proposed that all public sector procurers be subject to a new ambitious target that 50% food spend is on food produced locally or certified to higher environmental production standards. At the time of writing (July 2023) the outcome of the consultation has not yet been announced (DEFRA, 2022).

Proposed national contract, February 2023

In February 2023 the UK Cabinet Office announced that Crown Commercial Services, the government procurement agency, would set up a national contract for public sector food which would replace regional and local buying and distribution agreements set up by local authorities and the devolved administrations in Scotland and Wales. There was immediate vigorous criticism from local authorities that this had been announced with no consultation and it would destroy all the good work many local authorities had done in terms of sustainability, food standards, encouraging the use of British food, local supply and the adoption of SME's as suppliers (Foad, 2023).

A more recent notice on the Crown Commercial Services website indicates that participation by local authorities and the devolved administrations in the new national contract will be voluntary. Publication of the tender is scheduled for summer 2023 with the contract to be awarded by spring 2024 (Crown Commercial Services, 2023).

Denmark

Introduction

Denmark is one of Europe's major food exporting countries. This may explain why public procurement policy in Denmark has moved away from emphasising procurement of local food.

Between 2013 and 2016 the Danish government funded the *Kloge Fødevareindkøb* [Wise Food Procurement] scheme which encouraged municipalities to buy organic food and also to source from small local suppliers. But in November 2016 the national public procurement agency SKI took a decisive step in a different direction by awarding a national contract for public food. Nonetheless, some municipalities have continued to purchase at least some of their food requirements from local suppliers.

Kloge Fødevareindkøb [Wise Food Procurement]

Between 2013 and 2016 the *Kloge Fødevareindkøb* project received government funding to work with municipalities to improve public food procurement. The project was run by a partnership of two organisations: Madkulturen [food culture] and København Madhus [Copenhagen House of Food].

In November 2016 Kloge Fødevareindkøb produced extensive guidance for public purchasers on how they should try to encourage small local food suppliers through such means as:

- Direct procurement
- Simple tenders – to encourage small suppliers to submit proposals
- Division of tenders into lots to make things easier
- Dialogue with small suppliers so you can design the food procurement in such a way as to encourage them to bid
- Start the process well in advance – and be patient! (Kloge Fødevareindkøb, 2016.)

A detailed evaluation of the project's work was carried out by Aarhus University for the Ministry of Environment and Food and published in May 2017 (Pedersen & Jensen, 2017.) The report was based on 21 interviews with food suppliers, wholesalers, food hubs, municipalities and public institutions. Advice had been provided to 38 out of 98 municipalities, four out of five regions, five out of 13 purchasing communities and three governmental organisations. Counselling had focused on interpretation of existing supply contracts, the preparation of new ones and practical approaches to increase organic and local food.

All 98 Danish municipalities were offered advice about public food procurement but the majority were not interested. Municipalities experienced a number of barriers to increasing local/organic public food purchases. These included:

- Tender rules and procurement contracts
- Logistics for distributing food from small local producers
- Small suppliers unable to guarantee volumes required
- Suppliers reluctant to sell to the public sector
- Lack of political will in municipalities and kitchens

Of 23 respondents who received advice, 13 were satisfied with the advice. Of the five municipalities who were interviewed, three said that the advice received from the project had led to a successful outcome in terms of increasing organic and/or local food. The report suggested that the advice project should continue but no further government funding was made available. This reflected the change in government policy towards encouraging all municipalities to sign up to the national food contract

Introduction of the national contract

In November 2016 the national procurement organisation SKI awarded a national contract for food to a large food wholesaler, Hørkram. The national contract was joined by 65 of the country's total of 98 municipalities. Total contract value was around DKK 500 million per annum. Participating municipalities were however allowed to opt out of the national contract and make their own arrangements for specific products (Mørch, 2016).

The decision to set up this national contract was taken In 2014. It represented a radical change of course compared to previous government advice in 2013 encouraging public procurers to source from small local suppliers (Food Minister Mette Gjerskov quoted in Food Supply Denmark, 2013). The 2016 contract award coincided with new procurement legislation incorporating the provisions of the new EU procurement directive – which encouraged the breaking up of large contracts into lots. Objections to the national contract from small food suppliers were fruitless (Kongsgaard, 2018). On 31 May 2019 Hørkram was once again successful in winning the national contract (Kongsgaard, 2019). Most of the food supplied through the national contract – both organic and conventional – is Danish. Customers have praised Hørkram for high quality, good prices and service.

Local procurement: Case studies

Most Danish municipalities obtain their food through the national contract. However, certain municipalities have chosen to stay out of the national contract for at least some of their food requirements so as to source food from local suppliers.

Copenhagen

Copenhagen has sourced the great majority of its food from the national contract. However since it is the largest city in Denmark its local purchases were still quite sizeable. The City has documented their work on engaging local suppliers to help the City increase the percentage of organic food. The City talked to local producers before drawing up tenders to find out what these producers could supply and how tenders should be adapted to make it easier for these suppliers to supply the City. The City broke down contracts into lots to try to encourage smaller suppliers. Copenhagen is at present the only Danish municipality which is experimenting with using dynamic food procurement to buy food from local suppliers. Local suppliers deliver their food to the two large central kitchens operated by the City – one doing food for schools and the other doing food for elderly people (Bøttern, 2023). This minimises the logistical demands on smaller suppliers. The national contractor, Hørkram, delivers food to the many smaller kitchens within individual schools and nurseries.

Bornholm

Bornholm is one of the agriculturally most productive parts of Denmark. In 2007 the Regional Council adopted a sustainability strategy – Bright Green Island. – which envisaged that the municipality would have 60% organic food in public meals and 40% local food. Bornholm had a strong local network – municipality, food companies – large and small – working together with local farmers. A successful public campaign had saved the local abattoir from closure (Groth-Michelsen & Schenk, 2017). The increase in organic and local food required long-term political commitment and continual dialogue between all parties. The objective was to increase organic and local food while producing tasty meals within the same budget. Kitchen food requirements were analysed. Kitchen staff were retrained to cook from scratch – with support from Copenhagen House of Food. Suitable local suppliers were identified. A Bornholm food hub was established so that local products could be collected, brought to the warehouse and distributed to each kitchen (Madkulturen, 2016). By 2019 Bornholm had 20% local food in its public kitchens (Coff et al., 2020).

Ringkøbing-Skjern

This is a rural municipality with a large agricultural sector. The municipality has a longstanding policy of trying to assist small food producers to market their produce – including to public kitchens. A website has been set up www.smagenafvest.dk. It provides an overview of the municipality's food producers, farm shops and restaurants, with associated Facebook and Instagram accounts. The municipality has established a business park for small food producers, Food Park Skjern-Enge. The municipality has long planned to establish a food hub for local foods. In November 2021 these plans were fulfilled when Råhandel – now Pantrii – opened in the Food Park. Pantrii is an online marketplace where local food producers can sell their products to restaurants, specialty stores and/or selected Coop stores www.pantrii.io/. It has grown rapidly since it was originally established in Copenhagen in the summer of 2017 with a few producers and a warehouse in a garage for delivery and distribution of goods. Food Park Skjern-Enge has become the logistics hub for Pantrii covering the West Jutland area (Ringkøbing-Skjern Kommune, 2022).

Sønderborg

This municipality has produced long-term food strategies which focus on developing the local food sector. Most food has continued to be supplied through the national agreement but the following categories were to be bought outside the national agreement: fresh fruit and vegetables, fresh fish, baked goods, beverages and tea and coffee. Seasonal menus and cooking from scratch were advocated (Sønderborg, 2019). The municipality organised a meeting between buyers and small producers in August 2018. It was pointed out that breaking up the locale's procurement into smaller portions was an important way to help smaller producers win a share of the work. Opportunities for distribution collaboration were also discussed, as this is a barrier for small producers (Sønderborg Kommune, 2018). In March 2020 a delegation from Sønderborg visited Sweden to study coordinated distribution It continues to be the case that no municipality in Denmark has introduced coordinated goods distribution on the Swedish model.

In September 2022 the municipality approved a food action plan for the years 2022–2025 which continued to fund a food coordinator whose role included encouraging the purchasing of local food by the 20 public kitchens within the-municipality. Several municipal kitchens already had contracts with local bakers, butchers, fishmongers and greengrocers. The task was to ascertain the degree of success of these local arrangements and how difficult it would be for these suppliers to enter into agreements to supply all 20 kitchens. New potential suppliers would also be identified. The food coordinator was tasked with uncovering possibilities in relation to distribution solutions and arranging trials. The action would be a success when a sustainable distribution solution was found that enabled procurement from local producers to supply all 20 kitchens. The overall success criterion was that agreements would be established with local suppliers before the end of 2025 (Sønderborg Kommune, 2022).

Sweden

High food standards and increasing imports

In Sweden food imports have risen steeply since Sweden joined the EU in 1975. Swedish agriculture had been strictly protected from imports until the government decided that the whole economy required lower food prices to boost competitiveness. The increasing imports has driven down food prices but many Swedish farmers have found it very difficult to compete, particularly since Swedish agriculture has among the best animal welfare and environmental standards in the world. Many public procurers have bought a great deal of imported food. Others have focused on trying to buy locally to support local farmers – or Swedish food if what they require is not available locally.

Decentralised approach

Sweden has a decentralised approach. It has a large number of small municipalities – less than 20,000 population – which make their own procurement arrangements or form into small groups. The typical pattern is groups of between four and seven municipalities with total populations of less than 200,000

The majority of municipalities rely on the food supplied by two very large national wholesalers – Menigo and Martin & Servera, which have a national network of distribution centres and a dominant market position.

Legal disputes: Sigtuna judgment 2012

Historically public food procurement in Sweden has experienced a significant number of legal disputes, in which suppliers have succeeded in overturning procurement decisions made by public buyers, arguing that they were in breach of EU law. (Pedersen, 2011). The 2012 Sigtuna judgement was an important milestone. Sigtuna municipality had imposed animal welfare requirements within its meat tender which, among other things, restricted antibiotic use and transport times for slaughter. The large wholesaler Martin & Servera took legal proceedings, arguing that these requirements were unlawful. The court's decision that Sigtuna had acted lawfully was an important green light to encourage municipalities to impose requirements in tenders which in practice favoured Swedish food (Ander, 2015).

Procurement agency, September 2015

In September 2015 the newly-elected Social Democrat led government set up a new government department – the procurement authority (Upphandlingsmyndigheten) to provide advice and support for public buyers and suppliers. By giving authoritative advice to public buyers, the intention was to minimise further legal challenges to procurement decisions – including those aimed at encouraging Swedish suppliers – including smaller companies – to have a greater share in public procurement. The Civil Affairs Minister who brought about this change was Ahmed Shekarabi – previously the law professor who had advised Sigtuna Kommune in its successful lawsuit in 2012 (Ander, 2015).

The number of lawsuits mounted by Menigo and Martin & Servera which sought to "review" (overturn) local authority procurement decisions fell from 20% of all current agreements in 2015 to 12% in 2016 (Kost Och Naring, 2016). The Swedish Association of Public Purchasers reported in May 2021 that the number of such reviews had fallen further to around 5% of all tenders (Sveriges Offentliga Inköpare, 2021).

Techniques for encouraging local suppliers

The national procurement agency has published a brochure aimed at encouraging Swedish food producers to bid for public sector contracts. (Upphandlingsmyndigheten, 2019). On its website the agency has set out detailed guidance showing how public procurers can encourage local suppliers while staying within the limits of EU procurement law (Upphandlingsmyndigheten, 2021a, 2021b.) This should increase competition and push down prices. Procurers are advised to:

- Have a dialogue with potential smaller suppliers in their local area as part of their market analysis prior to issuing any tender
- Give smaller suppliers sufficient time to decide whether they want to bid for public sector business – and do not ask them to bid at harvest time
- Divide procurement into smaller parts (lots) whenever possible
- Consider logistical options for encouraging smaller suppliers – such as coordinated distribution (Upphandlingsmyndigheten, 2020; see later)

National food strategy, June 2017

The first Swedish National Food Strategy was approved by the Swedish Parliament with cross-party support in June 2017, setting out policy until 2030. The Strategy aimed at promoting greater food production within Sweden – both conventional and organic. Swedish conventional food is produced to high environmental and animal welfare standards. Small and medium-sized suppliers should be encouraged to tender for public food and catering (Regeringkansliet, 2017, pp. 3, 9, 11, 17; Ministry of Enterprise and Innovation, 2017). The National Food Strategy included a package of new financial assistance for the promotion of organic food in public catering and private restaurants. By the beginning of 2018 the government had formulated goals that by 2030 30% of Swedish agricultural land should be organic and public consumption in the same year would consist of 60% organic food (Koch et al., 2018).

Logistics: Coordinated distribution

In Sweden many municipalities have established coordinated goods distribution [*samordnad varudstribution*] projects of a type which is hardly found anywhere else in Europe. These are made possible by the large scale of public food provision in Sweden whereby municipalities are responsible for providing meals for children – from nurseries to secondary schools – and elderly people.

Suppliers of food or other products bring these to the distribution centre, where a contractor appointed by the municipality will be responsible for distribution to schools, nurseries and care homes. This should lead to fewer environmental impacts from road traffic and may also encourage small businesses to sell to the municipality (Bosona et al., 2013; Moen, 2014).

The National Centre for Coordinated Distribution received government funding between June 2018 and December 2020 to promote this approach (Nationellt centrum för kommunal samordnad varudistribution, 2019). By November 2020 the number of municipalities which had established coordinated goods distribution had risen to 45 (Holm, 2020).

Other municipalities have continued to make use of the large wholesalers and these wholesalers have lobbied against provision of public money to promote coordinated distribution – arguing that they already provide municipalities with an excellent service which minimises vehicle movements and supports local suppliers (Mårtensson, 2020b).

Percentages of Swedish beef and chicken in municipalities

The Swedish Food Agency has collected statistics showing the percentages of Swedish beef and chicken in each municipality (Livsmedelsverket, 2018). Sweden is unique in Europe for the public availability of detailed information about its public services.[2]

Sveko: Swedish organic % in municipalities

Organic food policies in Swedish municipalities have led to imports when Swedish organic food was not available. Since 2017 the Swedish Organic Food Centre has published a league table showing what percentage of purchases in each municipality was both organic and Swedish (Ekomatcentrum, 2021).

Local food procurement: Case studies

Within Sweden there are numerous municipalities which have made considerable efforts to source food from local suppliers. The following sections outline five examples. The first three all make use of coordinated distribution centres.

Växjö

Växjö promotes itself as being one of the greenest cities in Europe (Slavin, 2015). There is strong political support for sustainable public procurement policies. As regards food these included increasing the percentage of organic and locally-sourced food (from within a 15-km radius). Small companies from within the region were encouraged to participate in food tenders. The overall food contract was divided into smaller lots and the municipality signed agreements with all 17 companies who submitted bids. All suppliers delivered their food to a single distribution hub, rather than having to take it out to the 300 sites where the food is actually used. Of the meat purchased in 2018, 98% was Swedish (Braic, 2013; Växjö, 2019).

Trelleborg

Trelleborg announced in May 2023 that they had agreed a new, four-year food contract whereby the municipality gave greater opportunities to smaller suppliers. The new agreement had local suppliers for dairy products and baked products and meat was supplied by a regional farmers' association. All other food was provided by the national wholesaler, Martin & Servera. The contract also required an increase in the proportion of organic products from the current 23% to around 40%. The organic percentage was increased to 77% for fresh meat and 83% for dairy. The increased cost of organic food was funded by a 25% budget increase (Mårtensson, 2023).

Sodertalje

This municipality has taken a leadership role in European projects promoting healthy and sustainable food – particularly Diet for a Green Planet, Diet for a Clean Baltic and Baltic Ecological Recycling Agriculture & Society (BERAS). It took a political decision in 2006 to increase purchasing of organic and local food for public kitchens. Local food was defined as a maximum of 50-km distance from the municipal boundary (Grantstedt & Hertwig, 2015). The municipality has received European funding for the Matlust project which helps local suppliers develop new products for sale within public kitchens. The municipality has been active with share farming *(andelsjordbruk)* whereby small farmers enter into a contract with the municipality to produce food for public kitchens. For example the Under Tallarna farm producing vegetables for the Starrbäckens nursery (Lansstyrelsen (2021).

Skelleftea

This municipality has sought to use public meals as an educational tool. Every public kitchen was paired with a local farm. The goal was to spread knowledge about how food was produced and why it was important that there were food producers in the immediate area. Teachers in schools and preschools as well as staff in elderly care should take part in this learning together with children and the elderly. The school will go on a field trip to the farm and the kitchen will buy an animal from the farm and cook different dishes from it. The farmer might be present in the school restaurant when it served the meat from the farm. (Mattanken, 2023)

Ödeshög

Espen Rognstad of Ödeshög municipality described the adoption of a new way of procurement which increases the share of locally produced products. Ödeshög collaborated with three other locales – Vadstena Boxholm and Ydre – making up a total population of 24,000. A key objective was to increase the share of locally produced products. In the procurement they scaled down the number of products/lines in the request document from 1,500 to the approximately 350 largest products and stipulated that they wanted Swedish origin where possible.

> We asked for a conversion figure, that is, a percentage that the wholesaler adds to the supplier's price. This is also called open book procedure. The large wholesalers chose not to respond to this procurement, so the result was that we were able to conclude an agreement with a wholesaler based in the local area. The wholesaler is regionally anchored and has knowledge and existing contacts with local producers. Thanks to that, it has become easier for us to get products from local suppliers. . . . This has led to a local supplier who, among other things, grows cucumbers thinking about scaling up their business, now that the wholesaler they usually deliver to also has municipalities as customers.
>
> (Landsbygdsnatverket, 2023)

Espen also mentioned that the municipalities had been working for several years with a concept called Food for Life, which was inspired by contacts with British municipalities and observing their local food procurement policies.

> Within the concept, it is central to have contact with food producers, to gain knowledge about how our food is produced and to offer the opportunity for study visits. . . with this wholesale agreement, opportunities are opened up to further develop that way of working.

Dynamic food purchasing

At least 12 Swedish municipalities have started to use dynamic food purchasing. The examples which have so far been publicised are two groups of small municipalities:

- Tierps, Östhammars, Heby and Knivsta municipalities (combined population 88,000) started to use DIS for food procurement in March 2020 (Mårtensson, 2020a)
- Växjö, Alvesta, Lessebo, Ljungby, Markaryd, Uppvidinge and Älmhult (combined population 97,000) started in June 2022 (Växjö, 2022)

In both cases there was a clear aspiration to use the system to source more food from local suppliers.

Conclusions: Comparison of the three countries

There are marked differences between the three countries.

Differences of scale

Sweden has adopted a decentralised approach – with food procurement carried out typically by small groups of municipalities. This gives a great deal of scope for local initiatives including local food procurement.

Denmark has moved from encouraging local procurement to adopting a national food contract to which almost all municipalities subscribe, which leaves few opportunities for small and local suppliers.

The UK considered setting up a national contract in 2010 and this is once again under discussion at the time of writing (July 2023). Municipalities have set up relatively large-scale food procurement contracts which have very largely excluded small and local suppliers.

In Scotland and Wales the devolved governments have set up national food contracts – although municipalities are also given discretion to buy from small and local suppliers.

Techniques for promoting local supply

Rather similar techniques have been proposed to promote municipal purchasing from small and local suppliers – particularly analysis of food requirements, dialogue with small and local suppliers and breaking up of contracts whenever possible – so that they would be appropriate for smaller suppliers. Timing however in the three countries was very different.

In the UK the major effort to promote these techniques was in the time of the Public Sector Food Procurement Initiative – between 2003 and 2009.

In Denmark these techniques were promoted by a government-funded project between 2013 and 2016. However since the adoption of the national food contract in November 2016 this has provided food for the great majority of municipalities and only a relatively small number of municipalities have attempted to source food locally – including the examples given in this chapter.

In Sweden these techniques are widely followed. They have been promoted by national government since the establishment of a new national procurement agency (*Upphandlingsmyndigheten*) in 2015.

Logistics

In all three countries large national wholesalers have a dominant position in controlling the supply of food to municipalities. And small suppliers face very considerable difficulties with accessing the public procurement market because they do not have the resources to deliver their products to large numbers of public kitchens scattered around a municipal area.

Sweden is unique in Europe in that a minority of municipalities have set up coordinated good distribution centres which greatly reduce their dependence on the large national wholesalers and have considerable potential to reduce environmental impacts from vehicle movements and open up opportunities for smaller suppliers.

In Denmark the small number of municipalities adopting local procurement initiatives have all had to grapple with the issue of logistics.

Dynamic food procurement (DFP)

This has been proposed as a way in which municipalities can open up opportunities to small and local suppliers.

It is being used on a small scale in Denmark by Copenhagen City Council and by two groups of small municipalities in Sweden. It was used successfully by a single UK local authority, Bath & North East Somerset, between 2016 and 2018. In recent years throughout the UK there have been moves to promote DFP as a way to encourage sourcing of high-quality local food.

Notes

1 Lancashire County Council's contract has made some effort to open up opportunities to smaller suppliers to supply food to the County Council by requiring its fruit and vegetable wholesaler to provide a distribution service. This has led to the award of a separate contract for cheese supply. (Lancashire County Council, 2018). North Yorkshire County Council has attempted to encourage smaller suppliers by dividing its food contract into four lots for different parts of the county.

2 For discussion of the tradition of transparency in Swedish local government, see Kuhlmann (2010).

References

Ander, G (2015) *Shekarabi vill ha mer svenskproducerad mat I de offentliga köken* [Shekarabi Wants More Swedish-Produced Food in Public Kitchens]. Landlantbruk. 22 December 2015. www.landlantbruk.se/lantbruk/shekarabi-vill-ha-mer-svenskproducerad-mat-i-de-offentliga-koken/

Bonfield, P (2014) *A Plan for Public Procurement A Plan for Public Procurement Enabling a Healthy Future for Our People, Farmers and Food Producers*. https://assets.publishing.service.gov.uk/government/uploads/system/uploads/attachment_data/file/332756/food-plan-july-2014.pdf

Bosona, T, Nordmark, I, Gebresenbet, G, & Ljungberg, D (2013) GIS-Based Analysis of Integrated Food Distribution Network in Local Food Supply Chain. *International Journal of Business and Management*, 8(17), 13–34. https://doi.org/10.4236/jssm.2011.4202

Bøttern, E (2023) *Dynamic Purchasing System*. Presentation at Brussels Seminar, How Can Local Governments Support Small Farms? 22 March 2023. https://iclei-europe.org/calendar/?c=search&uid=shaMg6MY

Braic, D (2013) *Interview: The Strategic Importance of Green Procurement at Växjö*. GPP Issue No. 32 News-Alert June 2013. https://ec.europa.eu/environment/gpp/pdf/news_alert/Issue32_NewsAlert_Interview.pdf

Coff, C E, (Ed.), Ravn, M, Laugesen, S M B, Hansen, M W, & Christiansen, M (2020) *Rural Public Food Policies and Strategies: Case Studies From Denmark, Sweden and Lithuania* (1st ed.). Professionshøjskolen Absalon. www.ucviden.dk/ws/files/118566580/SBFOODINNO_Rural_Public_Food_Policies_and_Strategies_pdf

Crown Commercial Services (2020) *Food Supply and Delivery Platform*. 26 June 2020. www.youtube.com/watch?v=2WJ_3Xw-Img

Crown Commercial Services (2021) *Food Category Newsletter*. September 2021. https://assets.crowncommercial.gov.uk/wp-content/uploads/RM6279-September-2021-Newsletter-v1.pdf

Crown Commercial Services (2023) *Food and Drink*. 10 July 2023. www.crowncommercial.gov.uk/agreements/RM6279

DEFRA (2006) *Public Sector Food Procurement Initiative How to Increase Opportunities for Small and Localproducers When Aggregating Food Procurement Guidance for Buyers and Specifiers*. https://web.archive.org/web/20071003120000/www.defra.gov.uk/farm/policy/sustain/procurement/pdf/aggregation-guidance.pdf

DEFRA (2008a) *Public Sector Food Procurement Initiative Putting It into Practice Advice for Promoting Healthy Food and Improving the Sustainability and Efficiency of Food Procurement, Catering Services and Supply*. https://web.archive.org/web/20080620134824/www.defra.gov.uk/farm/policy/sustain/procurement/pdf/psfpi-putting-into-practice.pdf

DEFRA (2008b) *Case Studies Index*. https://web.archive.org/web/20081023175513/www.defra.gov.uk/farm/policy/sustain/procurement/casestudies/index.htm (Accessed 23 October 2008).

Defra (2022) *Public Sector Food and Catering Policy for England the Government Buying StandardsforFoodandCateringServices(GBSF)ProposedDocumentforJune2022Consultation*. https://consult.defra.gov.uk/public-sector-food-procurement/food-and-catering-consultation/supporting_documents/Proposed%20public%20sector%20food%20and%20catering%20policy%20for%20England%201.pdf

Deloitte (2009) *DEFRA-Public Sector Food Procurement Initiative an Evaluation-Final Report*. London: Department of Food & Rural Affairs. April 2009. www.defra.gov.uk/farm/policy/sustain/procurement/. https://web.archive.org/web/20090731145559/www.defra.gov.uk/farm/policy/sustain/procurement/pdf/090311-PSFPI-%20evaluation.pdf

Dynamic Procurement UK Food (2022) *The Dixon Foundation Match Funding Challenge*. www.dpukfood.co.uk/funding-opportunity

Ekomatcentrum (2021) *Marknadsrapport 2021 Ekologiskt I Offentlig Sektor 2020* [Market Report 2021 Organic in the Public Sector 2020]. http://ekomatcentrum.se/wp-content/uploads/2021/05/Rapport-Marknadsrapport-EMC-2021-PDFny-1.pdf

European Commission (2018) *Public Procurement: Legal Rules and Implementation*. http://ec.europa.eu/growth/single-market/public-procurement/rules-implementation_en

Farmgarden (2023) *Sustainable Food Procurement Hubs*. June 2023. www.farmgarden.org.uk/sustainable-food-procurement-hubs?utm_source=newsbrief&utm_medium=email&utm_campaign=June2023

Farming UK (2021) *Low Quality Food Imports Served in Schools, Charity Warns*. 21 April 2021. www.farminguk.com/news/low-quality-food-imports-served-in-schools-charity-warns_58038.html

Foad, D (2023) *Government Warned Its One-Stop-Shop Plan for Public Sector Food Could be 'Devastating'*. 15 February 2023. www.publicsectorcatering.co.uk/news/government-warned-its-one-stop-shop-plan-public-sector-food-could-be-devastating

Food Supply Denmark (2013) *Opfordring til kommunerne: Købocalee fødevarer* [Call for Municipalities: Buy Local Food]. 17 June 2013. www.food-supply.dk/article/view/105433/opfordring_til_kommunerne_kob_lokale_fodevarer

Foster, M (2018) *School Meals in North Yorkshire Get Top Marks for Freshness and Health*. 20 February 2018. https://www.thenorthernecho.co.uk/news/local/northyorkshire/16034174.school-meals-north-yorkshire-get-top-marks-freshness-health/

Goggins, G, & Rau, H (2016) Beyond Calorie Counting: Assessing the Sustainability of Food Provided for Public Consumption. *Journal of Cleaner Production*, 112, 257–266. https://doi.org/10.1016/j.jclepro.2015.06.035

Grantstedt, A, & Hertwig, J (2015) *Diet for a Clean Baltic in Practice*. https://urbact.eu/sites/default/files/import/Projects/Diet_for_Green_Planet/documents_media/DCB-cookbook_eng-webversion.pdf

Groth-Michelsen, L, & Schenk, L (2017) *Bornholm – Regional Transition Through Food Production*. Abstract for Sustain Conference, Danish Technical University. 6 December 2017. https://orbit.dtu.dk/ws/files/140585861/SustainAbstracts2017a.pdf

Hampshire County Council (2019) *Suppliers and Provenance*. https://www.hants.gov.uk/educationandlearning/hc3s/ourfood/suppliers

Harland, C, Telgen, J, Callender, G, Grimm, R, & Patrucco, A (2019) Implementing Government Policy in Supply Chains: An International Coproduction Study of Public Procurement. *Journal of Supply Chain Management*, 55(2), 6–25. https://doi.org/10.1111/jscm.12197

Holm, J (2020) *Kommunal samordnad varudistribution Skriftlig fråga 2020/21:479* [Municipal Coordinated Distribution of Goods Written Question 2020/21:479]. 13 November. www.riksdagen.se/sv/dokument-lagar/dokument/skriftlig-fraga/kommunal-samordnad-varudistribution_H811479

House of Commons (2020) *Written Evidence Submitted by Dynamic Food Procurement National Advisory Board (NAB) (PRO0003)*. https://committees.parliament.uk/writtenevidence/9762/pdf/

House of Commons Environment, Food and Rural Affairs Committee (2021) *Public Sector Procurement of Food Sixth Report of Session 2019–21 Report*. 3 March 2021. https://committees.parliament.uk/publications/5509/documents/54917/default/

Kersley, H (2011) *The Benefits of Procuring School Meals Through the Food for Life Partnership: An Economic Analysis for Food for Life Partnership*. London: New Economics Foundation. https://neweconomics.org/2011/05/benefits-procuring-school-meals-food-life-partnership

Kimbell, J (2023) *How will Public Procurement Change Post Brexit?* 3 January 2023. https://blog.insidegovernment.co.uk/en-gb/finance-and-procurement/how-will-public-procurement-change-post-brexit

Kloge Fødevareindkøb (2016) *Kvikguide for nye indkøbere* [Quick Guide for New Buyers]. http://wearecrunch.net/madkulturen/forside/kvikguide/ (Accessed 13 November 2016).

Koch, C, Ascard, J, Falkenek, A, Jönsson, A, André, C B, Fröman, E, Gotting, M, & Andresen, N (2018) *Åtgärdsplan för att öka produktion, konsumtion och export av ekologiska livsmedel. Rapport från ett regeringsuppdrag om ekologiska livsmedel* [Action Plan to Increase Production, Consumption and Export of Organic Food. Report from a Government Commission About Organic Foods]. https://www2.jordbruksverket.se/download/18.6412fcaf162e485c3c1b4ecf/1524546625771/ra18_16.pdf

Kongsgaard, H (2018) *Annulleret kæmpeudbud bør få store konsekvenser* [Canceled Giant Tender Should Have Major Consequences]. 15 February 2018. https://www.food-supply.dk/article/view/585825/annulleret_kaempeudbud_bor_fa_store_konsekvenser?rel=related

Kongsgaard, H (2019) *Storkøkkener: Hørkram beholder kæmpestort udbud* [Large Kitchens: Hørkram Retains Huge Supply]. 3 June 2019. https://www.food-supply.dk/article/view/664020/storkokkener_horkram_beholder_kaempestort_udbud?ref=newsletter&utm_medium=email&utm_source=newsletter&utm_campaign=daily

Kost Och Naring (2016) *Grossistbarometern 2016* [The Wholesaler Barometer 2016]. www.kostochnaring.se/vart-arbete/upphandling/grossistbarometern/

Kuhlmann, S (2010) Performance Measurement in European Local Governments: A Comparative Analysis of Reform Experiences in Great Britain, France, Sweden and Germany. *International Review of Administrative Sciences*, 76(2), 331–345. https://doi.org/10.1177/0020852310372050

Lancashire County Council (2018) *Procurement Title: Provision of a Food Distribution Network Procurement OptionOJEU – Open Tender Procedure*. http://council.lancashire.gov.uk/documents/s138768/Appendix%20A.pdf

Landsbygdsnatverket (2023) *Nytt sätt att upphandla ökar andelen närproducerat* [New Way of Procurement Increases the Share of Locally Produced Products]. 24 February. www.landsbygdsnatverket.se/mattanken/omoss/nyheter/nyhetsarkiv/nyttsattattupphandlaokarandelennarproducerat.5.63a7ae421865f604b9b7f10f.html

Lansstyrelsen (2021) *Avtal mellan andelsjordbruk och offentliga kök* [Agreement Between Cooperative Farming and Public Kitchen]. www.lansstyrelsen.se/download/18.3da1c377162bd90d9ee11e77/1526068463283/avtal-andelsjordbruk-offentliga-kok.pdf

Lawes-Johnson, A & Woodward, J (2022) *A Tale of Two Counties: Strengthening Local Food Cultures Through Mapping Supply Chains in East Sussex and Lancashire*. www.sustainweb.org/reports/mar22-a-tale-of-two-counties/

Livsmedelsverket (2018) *Fakta om offentliga måltider 2018 – resultat redovisade per kommun* [Facts About Public Meals 2018 – Results Reported by Municipality]. www.livsmedelsverket.se/maltidsfakta#Livsmedelsverkets%20kartl%C3%A4ggning%20av%20offentliga%20m%C3%A5ltider%202018

Madkulturen (2016) *Baggrundsviden om Bornholms Regionskommune* [Background Knowledge About Bornholm Regional Municipality]. https://web.archive.org/web/20171106002659/http://klogef%C3%B8devareindk%C3%B8b.dk/cases/bornholm/baggrund/

Mårtensson, F (2020a) *Inför DIS-modell för livsmedel* [Introducing DIS Model for Food]. 1 December. https://upphandling24.se/infor-dis-modell-for-livsmedel/

Mårtensson, F (2020b) *Hellre dialog än samordning* [Dialogue Rather Than Coordination]. 9 March 2020. https://upphandling24.se/hellre-dialog-an-samordning/

Mårtensson, F (2023) *Rivstartade nytt avtal med mässa* [Started a New Agreement with Messa]. https://inkopsradet.se/rivstartade-nytt-avtal-med-massa/

Martin, S, Hartley, K, & Cox, A (1999) Public Procurement Directives in the European Union: A Study of Local Authority Purchasing. *Public Administration*, 77(2), 387–406. https://doi.org/10.1111/1467-9299.00159

Mason, R (2022) *Flagship Post-Brexit Australia Trade Deal 'Not Actually Very Good', MPs Hear*. www.theguardian.com/politics/2022/nov/14/flagship-post-brexit-australia-trade-deal-not-actually-very-good-george-eustice

Mattanken (2023) *Nu har varje kök "en egen bondgård*" [Now Every Kitchen Has "Its Own Farm"]. https://mattanken.se/mattanken/erfarenhetsbank/erfarenhetsbank/nuharvarjekok enegenbondgard.5.3d5cb39018694b4f194486fb.html

McKevitt, D, & Davis, P (2015) How to Interact, When and With Whom? SMEs and Public Procurement. *Public Money & Management*, 35(1), 79–86. https://doi.org/10.1080/0954 0962.2015.986897

Michaels, S (2006) *Best Practice in Sustainable Public-Sector Food Procurement. Commissioned by Food Links UK, Funded by Defra*. June 2006. www.localfood.org.uk/library/ Defra-FLUK%20best%20practice%20final%20June%2006.pdf

Ministry of Enterprise and Innovation (2017) *A Long-Term Food Strategy for Sweden*. Press Release. EN. 6 April 2017. https://www.government.se/articles/2017/04/a-long-term-food-strategy-for-sweden/

Moen, O (2014) Co-Distribution of Municipal Goods in Sweden – Procurement From a New Standpoint. *Procedia-Social and Behavioral Sciences*, 125, 484–495. https://doi.org/10.1016/j.sbspro.2014.01.1490

Mørch, T (2016) *Udbudsbetingelserne for 'udbud 50.90 Fødevarer'* [Terms and Conditions for 'Tender 50.90 Food']. 5 December 2016. https://fodevarewatch.dk/Fodevarer/ article9202280.ece

Morgan, K, & Sonnino, R (2008) *The School Food Revolution: Public Food and the Challenge of Sustainable Development*. London: Earthscan.

Morley, A (2023) *Can Regional Food Systems Feed the Public Plate*? March 2023. www.foodforlife.org.uk/~/media/files/policyreports/adrain-morley_regional_food_system_ full-report.pdf

National Audit Office (2006) *Smarter Food Procurement in the Public Sector*. https:// webarchive.nationalarchives.gov.uk/ukgwa/20170207052351/www.nao.org.uk/wp-content/uploads/2006/03/0506963_Ies.pdf

National Farmers Union (2023) *Australia and New Zealand Trade Deals Come into Force*. 11 May 2023. www.nfuonline.com/updates-and-information/australia-and-new-zealand-trade-deals-come-into-force/

Nationellt Centrum för Kommunal Samordnad Varudistribution (2019) *Om Nationellt centrum för kommunal samordnad varudistribution* [About the National Centre for Coordinated Kommunal Distribution]. www.samordnadvarudistribution.com/om-nationellt-centrum/

Pedersen, A B, & Jensen, A (2017) *Evaluering af Kloge Fødevareindkøb. Lokale og økologiske offentligeødevareindkøbb* [Evaluation of Wise Food Procurement. Local and Organic Public Food Procurement]. Aarhus Universitet, DCE – Nationalt Center for Miljø og Energi, 90 s. – Videnskabelig rapport fra DCE – Nationalt Center for Miljø og Energi nr. 229. http://dce2.au.dk/pub/SR229.pdf

Pedersen, K (2011) Upphandlingskrönika – Livsmedelshänsyn vid livsmedelsupphandlingar [Procurement Chronicle – Animal Welfare in Food Procurement]. *Europarättslig tidskrift (European Union of Journalism)*, 787–792. www.delphi.se/uploads/2018/08/ upphandlingskronika42011-kristian-pedersen.pdf

Regeringkansliet (2017) *A National Food Strategy for Sweden – More Jobs and Sustainable Growth Throughout the Country*. Short Version of Government Bill 2016/17:104. EN. https://www.government.se/498282/contentassets/16ef73aaa6f74faab86ade5ef239b659/livsmedelsstrategin_kortversion_eng.pdf

Ringkøbing-Skjern Kommune (2022) *Erhvervs-, Klima- og Arbejdskraftudvalget. . . 3. Sådan går det med fødevarepolitikken* [Business, Climate and Workforce Committee. . . 3. This is How the Food Policy is Going]. 24 October. www.rksk.dk

Slavin, T (2015) *What Can the World Learn from Växjö, Europe's Self-Styled Greenest City?* 25 November 2015. https://www.theguardian.comgre/cities/2015/nov/25/what-can-the-world-learn-from-vaxjo-europes-self-styled-greenest-city

Social Farms & Gardens (2023) *The Role of Public Procurement in Transforming Wales' Food System*. May 2023. www.foodsensewales.org.uk/app/uploads/2023/06/BRO_SFG_Report_EMAIL_ENG.pdf

Soil Association (2019) *Food for Life Served Here Handbook 2019 – Schools*. https://www.soilassociation.org/media/20184/fflsh-standards-handbook-schools.pdf

Soil Association (2020a) *Shortening Supply Chains Roads to Regional Resilience*. www.soilassociation.org/media/20821/shortening_supply_chains_report_web.pdf

Soil Association (2020b) *House of Commons Environment, Food and Rural Affairs Committee: Public Sector Procurement of Food, Written Evidence Submitted by the Soil Association (PRO0001)*. https://committees.parliament.uk/writtenevidence/9028/pdf/

Soldi, R (2018) *Sustainable Public Procurement of Food*. European Committee of the Regions. https://cor.europa.eu/en/engage/studies/Documents/sustainable-public-procurement-food.pdf

Sønderborg (2019) *Handleplan for Fødevarer 2019–2021 Sønderborg Kommune som klimavenligt fødevareområde* [Food Action Plan 2019–2021 Sønderborg Municipality as a Climate-Friendly Food Area]. August 2019. https://sonderborgkommune.dk/sites/all/files/Forvaltninger/Intern%20Stab/Beredskab/Dokumenter/handleplan_for_foedevarer_2019-2021.pdf

Sønderborg Kommune (2018) *Lokale fødevareproducenter går sammen i netværk* [Local Food Producers Join Networks]. 25 April. https://sonderborgkommune.dk/aktuelt/arkiv/lokale-foedevareproducenter-gaar-sammen-i-netvaerk

Sønderborg Kommune (2022) *Handleplan for Fødevarer 2022–2025* [Action Plan for Food 2022–2025]. https://sonderborgkommune.dk/sites/default/files/2022-11/GOD-KENDT%20Handleplan%20for%20F%C3%B8devarer%202022-2025_0.pdf

SPP Regions Consortium (2017) *SPP Regions Tender Model – Food and Catering: Provision of School Food in Bath*. https://sppregions.eu/fileadmin/user_upload/Tenders/SPP_Regions_Tender_model_food_contract__final.pdf

Sustain (2021) *Leicestershire – Opening the Food and Climate Dialogue Across the County*. 4 October 2021. www.sustainweb.org/blogs/oct21-leicestershire-opening-the-food-dialogue/

Sveriges Offentliga Inköpare (2021) *Blandad kompott från livsmedelsfronten* [Mixed Compote from the Food Front]. 25 May 2021. www.soi.se/aktuellt/2021/maj/livsmedel-collage

Upphandlingsmyndigheten (2019) *Offentlig sector vill köpa svenskt* [The Public Sector Wants to Buy Swedish]. www.upphandlingsmyndigheten.se/globalassets/dokument/publikationer/broschyr-offentlig-sektor-vill-kopa-svenskt.pdf

Upphandlingsmyndigheten (2020) *Olika distributionslösningar vid livsmedelsupphandling* [Different Distribution Solutions for Food Procurement]. November 19. www.upphandlingsmyndigheten.se/branscher/upphandling-av-livsmedel-och-maltidstjanster/olika-distributionslosningar-vid-livsmedelsupphandling2/

Upphandlingsmyndigheten (2021a) *Underlätta för små och medelstora företag att lämna anbud* [Make It Easier for SMEs to Submit Tenders]. www.upphandlingsmyndigheten.se/inkopsprocessen/forbered-upphandling/sma-och-medelstora-foretag/

Upphandlingsmyndigheten (2021b) *Upphandling av närproducerade livsmedel* [Procurement of Locally Produced Food]. www.upphandlingsmyndigheten.se/branscher/upphandling-av-livsmedel-och-maltidstjanster/upphandling-av-narproducerade-livsmedel/

Växjö (2019) *Mat och hållbara måltider i skola och förskola* [Food and Sustainable Meals in School and Preschool]. 9 August 2019. https://vaxjo.se/sidor/forskola-och-skola/mat-och-hallbara-maltider.html

Växjö (2022) *Växjö kommuns satsning på småskaliga leverantörer bär frukt* [Växjö Municipality's Investment in Small-Scale Suppliers is Bearing Fruit]. 3 November 2022. www.vaxjo.se/sidor/jobb-och-karriar/nyheter – jobb-och-karriar/arkiv/2022–11–03-vaxjo-kommuns-satsning-pa-smaskaliga-leverantorer-bar-frukt.html

Wales Rural Network (2023) *Innovative Pilot Programmes Showcase Successful Models for Sustainable Food Procurement in Wales*. 16 June 2023. https://businesswales.gov.wales/walesruralnetwork/news-and-events/innovative-pilot-programmes-showcase-successful-models-sustainable-food-procurement

Wilkinson, T.1, Nye, C, Lobley, M., Clappison, A., Goodwin, A., Hilton, J., West, H., Thompson, S., MacQueen, G. and Fogerty, M. (2022) *Public Procurement of Food in the South West Region: opportunities and barriers to relocalising food supply*. Centre for Rural Policy Research, University of Exeter https://socialsciences.exeter.ac.uk/media/universityofexeter/research/microsites/centreforruralpolicyresearch/pdfs/researchreports/Public_Procurement_of_Food_in_the_South_West.pdf

4 Public meal provision in the Baltic Sea Region

Analysis of issues and challenges towards sustainable public food systems

Minna Mikkola, Urszula Ala-Karvia and Mark Stein

Introduction

This chapter analyses the challenges encountered by public procurement officers and caterers in five northern European countries when trying to move towards increasingly sustainable public meal provision. It was inspired by the experience of five countries working together within the EU Interreg Baltic Sea program project, StratKIT.

The full title of the StratKIT project was "Innovative Strategies for Public Catering: Sustainability Toolkit across Baltic Sea Region". It was funded by the EU Interreg programme between 2018 and 2021.

The project aimed to share best practice between public catering organisations in the countries involved. The results of the project's work have been reported in several publications such as Mikkola & Ala-Karvia (2020); StratKIT (2020); Hamann (2021); Ala-Karvia et al. (2022).

The chapter presents case studies from Denmark, Estonia, Finland, Poland and Russia, comparing approaches to sustainable public meals in these countries. With Poland and Russia, the locations of the case studies have been identified – as the City of Rybnik and the St. Petersburg urban region. With Denmark, Estonia and Finland, the location of the case studies is not identified.

The case studies describe the scope, financial arrangements and quality standards of public catering in each country, particularly school meals. This includes making reference to purchases of organic and local food and measures to reduce food waste and meat usage. They examined the organisation of public catering – particularly whether it was carried out by people directly employed by local authorities or by contractors and whether food was cooked in direct proximity to where it was consumed ("cook and serve") or in a central kitchen from which food was transported to the sites where it is eaten. Each case study included a site visit and consideration of issues and challenges.

Organisations participating in the project were asked to contribute written descriptions of their activities. The collection of information was conducted by a semi-structured survey, addressed to participating public procurement officers and catering managers. The survey sought to find out state-specific regulatory framing

DOI: 10.4324/9781003393023-6

of public meal provision and the municipal or site-specific implementation of operations as it was governed by local circumstances. The survey focused on food policies, local and organic food, nutrition recommendations, in-house versus contract catering services, financial arrangements for cost sharing, quality standards and preparation modes and waste management. The written answers were translated into English by the researchers and analysed to provide data for the StratKIT project reports.

The project organised site visits to selected public kitchens in each country whereby participants from other countries could see firsthand what was being done in specific kitchens. Site visits were to public kitchens identified as important and progressive food service organisations. Conferences were organised at country level and for all the Baltic countries to enable people to discuss how best practice was organised and delivered. The information collected was written up in project reports such as Mikkola & Ala-Karvia (2020) and StratKIT (2020).

Limitations

The project commenced work in April 2019 and the first project conference was held in Tallinn in November 2019. Between March 2020 and the end of 2021 project activities had to be conducted through online meetings and seminars due to the COVID-19 pandemic.

EU legal context

Denmark, Estonia, Finland and Poland are all within the European Union and therefore subject to EU laws relating to food hygiene, public procurement and organic food. As regards procurement, public authorities are obliged to put out to tender all contracts above a legally binding threshold amount and any EU supplier must be able to tender for this work. There is an EU legal framework for recognising food as organic – which has fostered the growth of organic food production and consumption.

The legal and regulatory framework relating to public meal provision in Russia is discussed in the Russia case study later.

Denmark

Scope, finance and quality

Denmark differs from other countries discussed in this chapter in the very limited scale of public catering provided for children. In 2017 it was estimated that 73% of Danish children ate packed lunches at school or in day care, 5% went home for lunch and 22% received their lunch from a canteen or catering arrangement (Landbrug & Fødevarer, 2017).

Individual municipalities have been able to decide whether or not to provide meals for children in schools and nurseries. There has been a gradual increase in school food provision. Aarhus, the second largest city – decided in 2021 that all schools in the city should be equipped with kitchens over ten years so that they would have the capability to provide school lunches – to be paid for by parents (Ruge, 2021). In most cases parents are expected to pay for school lunches, although there may be free school meal initiatives for at least some pupils in individual municipalities.

The overall standard of public sector food in Denmark is high – mandated by government guidance (Damsgaard, 2022). The Danish Food Authority awards the Healthy Meal label for public food which complies with the official nutritional guidelines.

Since 2015 national policies have encouraged increased organic food in public kitchens (Holmbeck, 2020). Individual municipalities have however retained discretion as to whether to introduce organic food into public kitchens and some – such as Sønderborg – prefer to prioritise local food (Sønderborg, 2019).

While some municipalities have little organic food in their public kitchens, others have very large amounts – over 80% in Aarhus and nearly 90% in Copenhagen. Usage of organic food in public kitchens in different areas can be tracked through the map of the Danish Organic Cuisine Label, whereby kitchens are awarded a gold, silver or bronze certificate according to the share of organic food in meals (www.oekologisk-spisemaerke.dk/omspisemaerket/danmarkskort/).

Denmark also has several retailers which have competitively promoted organic food items such as milk and supported the growth of a strong organic market. The situation also benefits public procurers because organic food is readily available and the organic price premium can be very modest (Olesen, 2017). Efforts to promote organic food in public kitchens have been linked to initiatives aimed at reducing food waste and promoting plant-based meals (Mikkelsen & Sylvest, 2012).

Organisation of public catering

The majority of public catering in Denmark is delivered by staff who are directly employed by municipalities, regions or other public organisations. In these cases, public procurement is engaged in sourcing food ingredients for these services. Since 2016 most Danish municipalities have purchased food through a national contract with a large national wholesaler, Hørkram, which offers both organic and conventional food and a high level of environmental and nutrition requirements. Some municipalities also have smaller scale arrangements for procuring certain foods from local suppliers.

Where the entire catering service is put out to tender, there are likely to be stringent quality requirements including provision of organic food, plant-based menus and waste reduction measures.

In Aarhus around one-third of the nurseries providing lunch used private catering contractors. The rest had their own employees cooking lunch. Some had fully sized kitchens enabling them to cook meals on site. Others received lunch already prepared from the central municipal kitchen or from an external contractor.

Some Danish municipalities have central kitchens from which food is delivered to the sites where it is consumed. Others have adopted the approach of locating kitchens on school or nursery premises – so that food is produced close to where it is eaten. The "cook and serve" approach.

Meals in schools and nurseries may have an educational purpose – children learning hygiene, good table manners, social competence, trying new foods and about where the food comes from. Children may be encouraged to help with cooking and given the opportunity to visit farms.

Site visit

This kitchen cooked and served hot meals for around 125 pre-school children and 25 municipal employees. Equipment was a small four-plate stove, two refrigerators and a tunnel dishwasher. The service was run by two employees, one cook with a bachelor's degree in nutrition science and a support worker employed on social grounds. From the cook's point of view, without the dishwashing and cleaning performed by the co-worker, the service could not be run. Additionally, there was the continuous concern of having no substitute if one of the two should fall ill.

While the municipal public procurement officer was responsible for the contract with a wholesaler which provided basic foods, the catering contractor also purchased fresh ingredients for this kitchen and two other small kitchens. These food items were bought at market prices, which fluctuated continuously, so the menu was adjusted every two weeks to emphasise whatever ingredients were cheapest at the time.

There was little written documentation of menus and nutritional values. The cooking of the meals very much depended on the experience and creativity of the cook. Meals consisted mainly of different tasty stews, skilfully cooked. They were based on various grains and, vegetables and a very limited amount of meat – three kilograms for 150 eaters. The cook had made a colourful, very presentable poster about the meals cooked, depicting the different dishes. The meals were put into large containers and taken to the children's rooms to be portioned out into individual bowls. The cook's nutritional education enabled her to evaluate the nutritional content of the meals. Since the catering service was dependent on a parental vote, parents frequently commented on the meals and particularly requested vegetarian and vegan food.

Issues and challenges

This report of the site visit illustrates that employees may have very pressured working lives.

Danish public meal provision aims at high sustainability goals, encouraging healthy, organic and plant-based food. There are well-known examples of good practice such as the cities of Copenhagen and Aarhus. However, the overall picture in Denmark is one of great local variations. There is only a limited and patchy school meal service and provision of meals in nurseries is not universal.

There are large local differences in the availability of organic and plant-based food. There is presently an ongoing political debate about widening availability of school meals and enlarging access to plant-based food in public kitchens for those who do not wish to eat meat (Madkulturen, 2022; Dansk Vegetarisk Forening, 2020).

Estonia

Scope, finance and quality

School lunches are available in all schools. The government subsidises school lunches in both public and private schools for primary and secondary school children (grades 1–12) by one euro starting in 2018. This subsidy regards only food costs. Municipalities may pay an additional subsidy to cover the cost of energy, water, equipment and labour but this is not the case everywhere in Estonia. Tallinn provides free school meals to all children while Tartu provides free meals for the youngest pupils (grades 1–4), while in grades 5–9 parents pay nearly one-third of the costs, and in grades 10–12 they pay more than half.

In nursery schools' public meals may be free or partly subsidised plus a parental contribution, depending on municipality.

The aim of providing school meals is to promote health and reduce social inequalities. School meal provision aligns with national health and well-being policy set out in the "Public Health Act" and its sub-regulation "Health protection requirements for schools and preschool childcare institutions" as well as "Health protection requirements for catering in health and welfare institutions". The text presents in detail the organisation of catering, food provision principles, energy and nutrient content of food, preparation of menus, presentation of nutritional information for customers and dining halls and the serving of meals (Estonia – Ministry of Social Affairs, 2010).

The school meal regulations specify that the school menus should be published at least ten days ahead. The ten-day menu should provide on the average 30–35% of the recommended daily nutrition for children. School meals should be varied, balanced, healthy and tasty. Further requirements included prohibition of high-fat fried meals, fruit at least three times a week, fish at least once a week, pasta no more than twice a week and whole grain flour.

There is quite a large amount of organic food produced in Estonia and there are organisations campaigning to promote organic food.[1]

There has been a gradual increase in the usage of organic food in public catering.

Since 2017, the government has administered a labelling system to designate caterers with different levels of organic food (20–50, 50–80 or 80–100% of ingredients are organic). The number of schools and nurseries which have notified the Agriculture and Food Board about the usage of organic food has increased significantly in recent years: three in 2018, 18 in 2019, 25 in 2020 and 46 (of which 37 qualified for an organic label) in 2021. The City Government of Tartu, Estonia's second city, has played a particular role in encouraging organic and by the end of 2021 31 school and nursery canteens were using organic food. (Vetemaa et al., 2022). This may have been assisted by strong organic agriculture around the town.

The higher cost of organic food has been an obstacle. A pilot project with five nursery schools in the City of Tallinn illustrates how caterers have redesigned the menu to introduce a minimum of 20% organic ingredients without a large increase in costs.

> Taking into account, for example, the availability and prices of local organic food. The benefits of organic food were explained to children, and they were taken to visit organic farmers. . . . Participants tried new products, reduced the amount of meat and used different cuts of meat, experimented with different salads and soups, and gave attention to the appearance and names of the food, which was important to children.
>
> (Stratkit, 2020, p. 62)

In recent years there has been growing awareness of the desirability of reducing food waste in public catering and introducing more plant-based foods (Haridus, 2022; Taimse Teisipäeva, 2021).

Organisation of public catering

Small and rural municipalities often run their own in-house public catering services. In large municipalities catering services are usually put out to tender to catering contractors. In some places the municipality procures services for all its respective institutions such as schools, awarding contracts to one or more caterers. Other municipalities have decentralised public procurement so that each public institution procures its services by itself. Headmasters of day care institutions and schools take on the role of public procurement officer although this often requires more competence and time than they have available.

Catering companies mainly serve "fresh meals", produced by the cook and serve mode. Sometimes a school has no space for cooking and prepared meals are delivered from a distant central kitchen. At one extreme a school on the western coast of Estonia receives meals from a central kitchen 90 km away.

Site visit

The modern school kitchen operated in a rather limited space and made healthy and tasty food including the main course, carbohydrate and vegetable side dish as well as fresh vegetables served with fresh dark bread, spread and milk. The food offer resembled the Finnish plate model and actually the catering company had previously had a Finnish parent from which it had separated. At the other side of the counter, the "dinner ladies" met the young people and portioned the hot dishes of the meals according to individual wishes. On the day of the visit the hot dishes were chicken with bechamel sauce and mashed potatoes. Above the counter on the wall the nutrition content of the standard meal was displayed in impressive detail – energy, protein, carbohydrate, fat and vitamin and mineral content.

After the counter children passed serving tables where they were free to take vegetables, bread and drinks. The school dining hall had a cheerful atmosphere with colourful melamine covered tables.

Issues and challenges

Caterers have sometimes had difficulty accommodating the higher price of organic food. This has particularly been the case with municipalities which have focused on seeking the lowest price for the catering contractor. Increasing plant-based food has been a way of financing more sustainable menus in countries such as Denmark and Finland, but this may arouse particular opposition in Estonia because Estonia is a meat loving society (Eesalu, 2022).

Finland

Scope, finance and quality

Finland has a long-established national system of free school meals, grounded in legislation (Lintukangas & Palojoki, 2016). This was originally introduced for primary school children and has now been extended to secondary schools. There is moreover extensive provision of meals to elderly people, university students and employees in many organisations.

Public meals are expected to comply with national nutrition recommendations which are themselves based on the Nordic Nutrition Recommendations. There are specific recommendations for babies, children, adults, the elderly, allergies and vegan diets.

The Finnish plate model envisages the ideal meal where one-quarter of the plate area belongs to a protein source – such as meat, fish or egg. Half of the plate is filled with cooked and raw vegetables. One-quarter of the plate is reserved for potatoes, rice and pasta (Tikkanen & Urho, 2009).

There is strong emphasis in public catering on measuring the nutritional content of menus and documenting that the correct nutritional levels are being achieved – using software for recipe planning and nutrition analysis.

There is also a strong emphasis on sustainability policies in public meal provision – with each municipality developing its environmental strategies. Key themes have included promotion of organic and local food and this has frequently been financed through reducing usage of expensive meat (Nuutila et al., 2018). In September 2020 the government announced a goal of 25% organic ingredients in public kitchens by 2030. EkoCentria is a long-established training and advice project which has been working with public kitchens to encourage them to buy more organic food through the Steps to Organic Program (Luomuravintola, 2023).

Organisation of public catering

Municipal public meal provision is divided between in-house provision and contract catering. Historically, the public meal has been provided in-house. And this is still typical, for smaller and rural municipalities. Lately, the large municipal catering organisations have often been reorganised as municipally owned catering companies. The new companies often have high-level positions for economic, environmental, nutrition and customer management and extensive staff. Some large municipalities run both catering modes. The in-house catering organisation has its dedicated share of catering sites, while the municipally owned catering company tenders to provide public meals for the remaining sites, in competition with private sector caterers. This mode of municipal arrangements allows the comparison of differently organised procurement and catering services in terms of costs and efficiency. To benefit from scale economies, smaller municipalities and hospital districts may pool their food procurement, whereby a lead manager takes care of tendering and contracting on others' behalf. Other municipalities prefer to retain control of their procurement with a view to maximising purchases from local food suppliers.

Limited budgets mean that public kitchens have to minimise their costs whenever possible. The hunt for low costs has its largest impact not on the quality of ingredients, the majority of which are sourced from within Finland, but on centralising kitchens with consequent distribution of meals to a network of food service sites. This becomes visible in central kitchens cooking daily meals from around several hundreds to thousands to more than one hundred thousand per day. These kitchens often deploy "cook-chill" methods, while smaller kitchens still rely on the "cook and serve" mode of fresh meal delivery.[2]

Site visits

Site 1: This is a municipally owned catering company. It was previously the municipal catering department and then it was separated off as a catering company which has to compete against other caterers for municipal catering business The company produced meals in a central kitchen, which was called a "food factory". This was a secure food production unit with strict entry supervision and a fleet of delivery trucks on the yard. It was a well-organised, hygienic, safe and efficient production unit.

The factory produced about 20,000 meals per day through the cook-chill mode and served about 80 catering sites. This was a highly efficient cooking process in terms of the number of meals per employee and the low-cost, high-volume meal transport. The company also managed tens of small kitchens which cooked meals from fresh ingredients.

Meals were provided in strict conformity to requirements set out in the tenders, which included very high nutrition and sustainability standards. The meal service had two options daily, one animal-based and one vegetarian. Some schools also required one vegan day a week. The kitchen used organic and Fairtrade products.

It sought to increase use of domestic fish and used legumes to replace meat when possible. The company also collaborated proactively with a start-up developing insect-based ingredients.

There was a five weeks rotating menu which was visible on the company web pages. This allowed for annual estimates of food volumes and costs.

The company used nutritional software which detailed the quantity of ingredients, their current prices and nutritional values in terms of energy, proteins, fats, vitamins and minerals. The software allowed continuous follow-up and adjustments in terms of changing costs of ingredients and varying availability of different foods from the wholesalers. There was particular emphasis on compliance with nutrition recommendations.

The meal service heavily emphasised avoidance of serving waste and plate waste at its buffet serving counters, challenging customers to "eat what they take and take what they eat". Leftover meals were daily sold at a cheap price for customers living near the catering sites. Waste management also involved recycling of biowaste, glass, metal, cardboard, and plastics.

The company encouraged plant-based consumption, they organised cooking classes and innovative recipe competitions for teenagers. It was recognised that parents needed to be given more understanding about what their children were eating, and it was important to convince them of the value of the company's meal service.

The company had a highly professional leadership structure. It was very aware that every two or three years it would need to tender once again for its contract. It carefully monitored meals uptake and customer satisfaction. There was a concerning trend for children to shun the free school meal in favour of food bought from nearby supermarkets.

Site 2: This large municipally owned catering company was set up through merger of municipal in-house units from several small municipalities. It had set up a new food factory which was well equipped and highly efficient. It produced daily meals for 7,000 customers with only four staff. It used an innovative process developed in Finland called "cook cold" whereby the meal is assembled in the food factory without being cooked. It is transported in refrigerated conditions to the site kitchens where it is cooked This approach uses much less energy than cook-chill where food is cooked in the central kitchen then cooled in a blast chiller and later warmed up in a site kitchen. Quality has been satisfactory. Recipes and ingredients have to be chosen which are suitable for this process (Filppa & Kotilainen, 2012).

Site 3: This was a smaller central kitchen which belonged to a small municipality. It produced about 2,500 meals per day using cook and serve mode, which were then delivered to dining rooms around the municipality. This kitchen made considerable use of organic ingredients – even very expensive organic meat. The procurer-caterer collaborated closely with start-up food industries in organic products. They aimed to produce better than average quality meals, catering for an upper middle-class population within their municipal area.

Issues and challenges

The scale and efficiency of the aforementioned food factories was built upon availability of large volumes of uniform quality ingredients from large-scale food processors. Food factories had difficulty including local small producers as suppliers. On the other hand, these large centralised food production units could offer large possibilities for growth, if small local meat, fish and vegetable companies do manage to start supplying them. There is an ongoing effort of persuading cooks that organic food is of superior quality and worth the greater initial cost. While the push for plant-based meals has been quite strong there is also reluctance to offend important local animal-based industry interests by reducing purchases of meat and dairy.

Poland (Rybnik)

Scope, finance and quality

In Poland the main category of public catering is school meals.

As of September 2022, primary schools have a statutory obligation to provide meals to students. According to Art. 106a sec. 1 of the Education Law Act, a primary school is obliged to provide students with one hot meal a day and give them the opportunity to eat it during their stay at school. School meals have to take into account government recommendations on the proper nutrition of children and adolescents.[3] The State Sanitary Inspection Department has responsibility for assessing the quality of public catering in schools and nurseries, particularly as regards food safety.

Participation in school dining is voluntary and school meals are only partially subsidised. The price which parents pay for meals is determined by the school principal and governing body. The price is calculated to cover the cost of the ingredients used to cook the school meals. It does not include employees' salaries and the costs of maintaining the canteen, which are covered by public funds. Where a school does not have a kitchen on its premises and school meals are delivered from a different school, the transportation fees are covered by the school. The school governing body may exempt parents fully or partially from payments for meals in the event of a difficult family financial situation.[4]

In 2018, a long-term government programme "*Meal at School and at Home*" [*Posiłek w szkole i w domu*] was established for the period 2019–2023 to fund local government social support activities such as assistance in the form of meals or cash benefit for the purchase of food.[5]

Several local authorities have experimented with introducing organic food into school meals – usually produced by local farmers (City of Tarnow – Ogrodinfo, 2019; City of Łomża – Tyczyńska, 2016; Subcarpathia Province – Wasak, 2023).

There have been considerable efforts aimed at reducing food waste in school kitchens (Global Survey of School Meal Programs, 2021).

Efforts to promote plant-based food in school kitchens are at a comparatively early stage. The international NGO Pro Veg commenced activities in Poland during 2020 (Proveg, 2020).

Organisation of public catering

In Poland school meals are most often provided through in-house catering by schools where the employees are directly employed by the school or the municipality.

A national survey of school meals showed that those prepared by a catering company were, on average, significantly more expensive than meals prepared in the school's own canteen and pupils' parents were more likely to purchase lunches in school when the meals were prepared in on-site kitchens (Global Survey of School Meal Programs, 2021).

The schools and nurseries employ catering managers who are responsible for preparing meals, developing menus which meet with nutritional requirements, ordering food products and managing food safety.

Private shops on school premises are still common in Poland. These shops are also covered by the Health Ministry regulation of 26 July 2016 which imposes limits on sugar, fat and salt content of foods provided. The same regulation covers vending machines, which many schools are phasing out.

Site visit

A representative of the Finnish project team visited two sites School Complex no. 3 and School and Pre-school Complex no. 14 in Rybnik in autumn 2021.

The on-site kitchens belonged to the municipality and the number of staff employed is fixed by the municipality depending on the number of pupils served. Both cooked food on the school site in a kitchen next to the dining room – examples of the "cook and serve" model,

Common practice in Poland is that kitchen and dining facilities are kept segregated, and the only contact between pupils and the canteen staff may be through a small window where meals are picked up. School and pre-school complex no. 14 in Rybnik had partially opened up the wall dividing the kitchen and canteen so that pupils could see the kitchen in operation. Both pupils and staff were very enthusiastic about the change.

Complex no. 14 imitated nordic examples by introducing fruit buffets for desserts allowing pupils to pick the fruits they like and decide quantity, which has significantly reduced food waste. It also introduced a waterless cleaning method using moist micro-fibre cloths inspired by Finnish canteens.

The kitchen managers were actively engaged in finding ways to minimise food waste and increase purchases of organic and local food and promote plant-based menus. In making these changes they had strong support from the City of Rybnik management.

Meal preparation in these kitchens was heavily dependent on the knowledge and experience of the cooks, making best use of available ingredients. There was little reliance on written documentation of menus. Nor was there much documentation showing the nutritional values of food provided.

Issues and challenges

School kitchen's employees often lacked training in sustainability and due to their long working careers were resistant to change and unwilling to experiment with the food.

With being segregated from the pupils and with high levels of food waste the kitchen employees often felt underestimated and disrespected.

Expenditure on additional equipment for the canteen was given low priority by school management and there was a long wait even for small items.

Russia (St. Petersburg)

Scope, finance and quality

St Petersburg is the second largest city in Russia, with a population of about 5.4 million people. There are about 2,300 public institutions, serving more than one million public meals daily – in educational and healthcare institutions and youth summer camps.

The city's department of Social Nutrition regulates public meal provision in the city and exercised detailed control over standards.[6] Nutritional values for protein, fat, carbohydrates, vitamins and minerals are specified for different types of customers. The department has devised rotating menus – incorporating detailed recipe cards – for different institutions for a 10-day cycle and a 70-page list of permitted food items. All catering contractors serving public institutions within the city are obliged to adhere to these menus and follow the recipe cards.

Free school meals were provided to young people from low-income families and also families with three or more children, plus orphans and disabled children. Partly subsidised meals were provided for young people in sport classes or cadet corps or those with certain, health problems. Part subsidy covered 70% of the meal price, with parents paying the balance. Moreover, all elementary school children received a 70% subsidised school breakfast.

Russian government policy strongly discourages food imports – so all food was sourced from within Russia. Catering contractors have expressed interest in purchasing more food from within the region because they thought quality might be better. However, local farmers have hitherto not been ready to enter the market because they did not know the specific regulations and norms concerning public sector catering services. The Council of St. Petersburg municipalities has expressed interest in promoting a dialogue with local food suppliers to encourage them to consider supplying public kitchens in future. Topics discussed may include procurement bureaucracy for farmers, as well as availability, processing and packaging issues of food items for caterers.

Little or no organic food was purchased for public institutions and this situation was not expected to change. There was no unified mandatory certification for organic food and farming in Russia. Small quantities of food were available which were self-certified by farmers as organic – particularly milk and vegetables.[7]

One local NGO – Animal Voice was engaged in promoting vegetarian food in schools (https://voicesforanimals.ru/food/ecology/). School caterers showed tentative interest in increasing plant-based food. But up till now few had done much to reduce meat or dairy usage in school meals. Centralised high-level rulings from the city government dictated the menus' ingredients in detail and made it very difficult to reduce meat or dairy. School administrators strongly supported continued provision of meat in school meals. It was possible that school meals delivered good quality proteins to young people who would otherwise have been undernourished.

Organisation of public catering

Public catering was delivered by catering companies under contract to the public sector. Schools purchased the services of catering companies using a digital platform (zakupki.gov.ru), after the district administrations of St. Petersburg had checked the tender documentation. The cheapest bid against the approved tender generally won the digital auction. Contracts were awarded for three years. Tender documentation focused on price – with very little attention paid to sustainability considerations.[8]

The catering companies active in the school catering market had to meet criteria prescribed by the Federal Service for Surveillance on Consumer Rights Protection and Human Wellbeing (Rospotrebnadzor www.rospotrebnadzor.ru/en/). The aim of these criteria was to only allow professional companies with food technological and analytic experience and competencies into the market.

The St. Petersburg Department of Social Nutrition issued 10-day rotating menus for social school meals (partly subsidised) to be served across the city by all contracted catering companies. As an example, lunch could include salad or vegetables, soup (as first course), meat, fish or chicken and a side dish such as pasta or potatoes (as main course) and rye-wheat bread and black tea with sugar.

However, if a catering company developed its own menu which received Rospotrebnadzor's approval, it could be implemented for social meals instead of the one dictated by the department.

There was a two-tier system of school meals in St. Petersburg. The subsidised social meals were strongly regulated by the city government. But pupils from better off families could buy from the commercial menu ("menu of free choice") which was not under any governmental regulation and provided more expensive dishes provided by caterers on a basis of supply and demand.

Site visit

The modern multistorey school was notably well equipped with two swimming pools, a theatre, a dance hall and at the entrance, a wall-to-wall glass cabinet full of glittering sports trophies. The dining hall and a very large professional kitchen were located at the basement. The kitchen had modern equipment and was well able to provide freshly cooked meals ("cook and serve") for more than 1,600 pupils and their teachers. The yellow tiled floors and colourful round tables made the big canteen an inviting lunch environment. The kitchen and the dining hall workforce was exclusively female, mainly elderly, hygienically dressed, serving young people amid a cheerful atmosphere. The older pupils picked up their meals from a buffet. For the younger pupils meals were served at the tables from a catering car. The tables were set ready with the drink, and apple juice seemed like the staple. On the day of the visit the meal included typical northern European food, such as fish patties, potato croquettes and salad of grated swedes and lingonberries.

Issues and challenges

There has been little interest hitherto in reducing food waste or meat usage. The issue of climate impact is hardly discussed. Organic food is virtually absent from the Russian market and its introduction in public kitchens is not a topic often discussed.

The school administrations were keen to improve the school catering services but had limited ability to influence arrangements put in place by the city administrators and the catering contractors. Rather more tasty and varied food would be welcome, instead of 'always pasta'. There was support for meals being largely meat-based and little interest in more plant-based food.

Caterers did have a small amount of scope to make suggestions to the school administrators with regard to increasing local food or serving more plant-based meals. There was however considerable uncertainty about their future in the public meals market because the city government was known to be considering award of all public meal provision to a single catering contractor.

Russia passed national legislation in March 2020 whereby free hot meals were to be provided for primary school pupils (ages 6–13 years). Implementation of this legislation in St Petersburg is expected in forthcoming years.

Conclusions

This section draws together common themes found in all the case studies and compares and contrasts the different countries.

Payment school meals

There are quite different practices in different countries.

Universal free school meals are the longstanding practice in Finland. In Estonia the government subsidises school lunches by one euro – intended to cover food costs. Municipalities may pay an additional subsidy to cover the cost of energy, water, equipment and labour but this is not the case everywhere – so there are municipalities where parents are required to pay a contribution. In Poland parents are expected to pay for the cost of the food used in school meals but the staff and kitchens are publicly funded and there is financial assistance for low-income families, at the discretion of the school governing body. In Denmark most schools expect children to bring packed lunches from home. The minority of schools which do provide meals may or may do so with some degree of public subsidy, but this is within the discretion of each municipality.

In Russia parents have up until now been expected to pay for school meals, but there has been financial assistance for deprived families. The Russian government has now announced the introduction of universal free school meals and implementation is awaited. Moreover, in St Petersburg there was a two-tier system whereby subsidised social meals were regulated by the city government. But pupils with money could choose from an expensive, unregulated menu.

Organic food

Increasing use of organic food has been a widespread theme within public catering in the four EU countries covered in this chapter. In Denmark the government is encouraging a level of 60% organic food in public kitchens, but individual municipalities have also had considerable discretion on whether to introduce high levels of organic food. In Finland the government has announced a goal of 25% organic ingredients in public kitchens by 2030 There has also been a recent increase in organic food in Poland and Estonia. The outlier here is Russia, where there is very little availability of food recognised as organic and public procurement has not moved to increase purchases of organic food.

Local food

Aspirations by at least some procurement people to buy more food from local suppliers was a common theme in all five countries. The case studies touched on some of the obstacles to local food procurement – such as the introduction of a national food contract in Denmark, the large scale of Finnish factory kitchens and Russian food suppliers lacking knowledge of how to respond to public procurement requirements.

Plant-based food and alternative proteins

Growing use of plant-based foods has been a widespread theme within public catering. It has been particularly well developed in Denmark. It has also grown rapidly in Finland and more recently and to a lesser extent in Poland and Estonia. The exception here has been Russia where the case study showed little support for increases in plant-based foods within school catering.

At the first Finnish site visit the company was working with a start-up developing insect-based ingredients.

Food waste

In Denmark, Estonia, Finland and Poland there have been efforts to reduce food waste in public catering. The approach shown in the Russian case study is different in that there appeared to be no policy of measuring or reducing food waste in the public kitchen.

In-house services versus catering contractors

Three of the site visits illustrated the delivery of public catering services by catering contractors – Russia, Denmark and Estonia. Poland and the three Finnish site visits illustrated catering services delivered by staff directly employed by municipalities.

Catering practices

Of the kitchens described in the site visits two of the Finnish kitchens illustrated the large-scale central kitchens which are widespread within Finland. Site 1 used cook-chill, and Site 2 used "cook cold". These two kitchens had low unit production costs and produced food of reasonable quality.

The kitchens described in the site visits in Denmark, Estonia, Poland and Russia all used the traditional "cook and serve" model which is predominant in these countries and so did Site 3 in Finland, which was a smaller central kitchen.

The same highly structured and intensively documented approach to menus could be seen in Finland and Estonia and also in Russia.

There was a strong contrast with approach taken in the Danish site visited. Here the highly qualified cook used her wealth of experience and creativity to produce tasty and nourishing meals with whatever ingredients were available and without substantial use of written documentation of recipes or documenting compliance with nutritional standards. In Poland the site visit showed that the approach involved heavy reliance on cooks' knowledge and experience to create meals with little written documentation of recipes or nutritional quality.

In those kitchens which were located close to the dining rooms where the food was eaten – in Denmark, Estonia, Poland and Russia – there were greater opportunities for a dialogue between the cooks and the children eating the food which could make for more satisfied customers and less food waste. The clearest evidence for this happening was in Poland where the two schools had partially opened up the walls dividing the kitchen and canteen so that pupils could see the kitchen in operation.

Notes

1 For leading organisations supporting organic agriculture in Estonia see the websites of the, Estonian Organic Agriculture Foundation www.maheklubi.ee and Organic Estonia https://organicestonia.ee/. The Organic Center of the Estonian University of Life Sciences is based in Tartu. www.maheklubi.ee/koostookogu/emu_mahekeskus/

2 The phrase "cook and serve" is commonly used for traditional catering operations where food is cooked in a kitchen adjacent to the dining room where it is served (Lee-Ross, 1999). "Cook and chill" refers to the practice of prepared meals being chilled after cooking to refrigerator temperature (3–5 centigrade) when they can be stored for a number of days and transported to a number of possible dining sites. "Cook and freeze" refers to prepared meals being frozen to approximately -20 to -30 degrees centigrade which ensures a much longer storage life with possible transport to numerous dining sites (Elansari &

Bekhit, 2014). The term "cook and hold" is sometimes used to refer to situations in which food is cooked and then kept warm for a period of time before it is served (Maietta & Gorgitano, 2016). It may be transported from the kitchen where it is cooked to several dining rooms using insulated containers.

3 A Regulation of the Minister of Health of 26 July 2016 defined the groups of foodstuffs intended for sale to children in educational system units and the requirements to be met by foodstuffs used in as part of public catering for children and youth. The recommendations regarding school and nursery school meals include more servings of vegetables and fruits and wholegrain cereal products, more oily fish and providing children with drinks with no more than a certain sugar content (Poland – Health Ministry (2016).

4 Full subsidies are granted for families being cared for by local social care institutions (These families also receive support during holidays).

5 Such assistance targets children, youth (module 1), and adults (module 2) who meet the conditions for receiving assistance specified in the Act on Social Assistance and the income criterion, i.e. 150%. income criterion entitling to social assistance benefits, as well as schools for administration of canteens and eating places (module 3). The Ministry of Family and Social Policy is responsible for the implementation of modules 1 and 2, while the Ministry of Education and Science is responsible for module 3. In late 2022, the Council of Ministers adopted a Resolution amending the programme. The changes concern the increase of the income criterion entitling to the use of support from 150% up to 200% percent criterion (referred to in Article 8 of the Social Welfare Act). As a result, support under the Meal at School and Home programme is available to people on low incomes, whose income does not exceed: PLN 1,552 – for a person living in single household (previously, it was PLN 1,164), PLN 1,200 – for a person in the family (previously PLN 900).

6 Saint Petersburg Government Decree (2004) No 296 "On the Department of Social Nutrition" defines the responsibilities and powers of this department which regulates public meal provision in the city, www.gov.spb.ru/gov/otrasl/socpit; www.socpitanie.spb.ru. The Law "On Social Nutrition in Saint-Petersburg", issued in 2019 defines terminology and standards for public meals.

The Social Code of St Petersburg published in 2019 sets out the categories of children granted full or partly subsidised school meals. Saint Petersburg Government Decree (2018) No 953 fixes the cost of public meals in educational institutes.

The menus for social meals for ten days are published on the official website of the Department of Social Nutrition of Saint Petersburg, www.gov.spb.ru/gov/otrasl/socpit/normativdoc/menyu/menyu-dlya-gbdou.

The Expert Coordination Council for the Development of Social and Healthy Nutrition in St Petersburg issues recommendations for healthy eating based on scientific research. Training on social nutrition' is provided by the training industrial complex on social nutrition of Sankt Petersburg (http://ukkusp.ru). Other important bodies dealing with public meals include the City Public Procurement Committee www.gov.spb.ru/gov/otrasl/kgz/ and the environmental protection and sustainable consumption Committee.

The Information and Methodological Center of St Petersburg [*Информационно-методический центр осущест*] offers public procurement training www.imc.spb.ru.

7 Within the Leningrad Oblast there were then five self-identified organic farms with the area of less than 100 hectares. Creating a Russia-wide or regional organic label for commercial purposes has been delayed due to lack of consensus about required production methods. Organic status has been certified through voluntary schemes such as Demeter or Global Ecolabelling Network's Internationally Coordinated Ecolabelling System (GEN-ICES, The Russian National Organic Union (Национальный органический союз) is working to develop an organic market in Russia http://rosorganic.ru. A new Federal Law on Organic Production of 25 July 2018 entered into force in 2020. Three organic food retailers operate in St Petersburg – VkusVill, Organica and Eco-market 'Health Compass'. These sell certified organic products produced in Russia.

8 Public procurement within St Petersburg complies with national laws on public procurement (Federal Law no 44-FZ "On the Contract System for State and Municipal Procurement of Goods, Work and Services" and Federal Law No. 223-FZ "On purchasing Goods, Work, and Services by Certain Types of Legal Entities").

The Federal Law No 44-FZ lists methods to determine the supplier by competitive methods such as open bids, limited participation bids, two-stage bids, closed bids, digital auctions and closed auctions. It is also possible to make purchases from a sole supplier.

These laws do not give weight to environmental considerations unlike the Green Public Procurement recommendations which influence procurement in the EU. While not absolutely excluded, environmental considerations are not regarded as a strategic issue.

References

Ala-Karvia, U., Góralska-Walczak, R., Piirsalu, E., Filippova, E., Kazimierczak, R., Post, A., . . . & Mikkola, M. (2022). COVID-19 driven adaptations in the provision of school meals in the Baltic Sea Region. *Frontiers in Sustainable Food Systems*, *5*, 750598. doi: 10.3389/fsufs.2021.750598

Damsgaard, L. M. (2022). *Kostråd til måltider: Nye anbefalinger til maden i skoler, daginstitutioner og kantiner* [Dietary Advice for Meals: New Recommendations for Food in Schools, Day Care Centers and Canteens]. Retrieved from https://goderaavarer.dk/nyheder/2022/kostraad-til-maaltider-nye-anbefalinger-til-maden-i-skoler-daginstitutioner-og-kantiner

Dansk Vegetarisk Forening. (2020, 28 December). *Kampen for en grøn valgmulighed i de offentlige køkkener* [The Battle for a Green Option in Public Kitchens]. Retrieved 2 January, 2021, from https://vegetarisk.dk/kampen-for-en-groen-valgmulighed-fortsaetter/

Eesalu, M. (2022, 8 April). *Plant Based Food in Estonian Schools – Is It Even Possible?* Presentation Given at Online Seminar of CLIKIS Project "Climate Friendly School Nutrition". Retrieved from https://drive.google.com/drive/folders/1fKLSgWVhtxxmG4LWnypz9BNswjtJoSXT

Elansari, A., & Bekhit, A. E. D. A. (2014). Processing, storage and quality of cook-chill or cook-freeze foods. In Rahman, M. S., & Siddiqui, M. W. (eds.). *Minimally Processed Foods: Technologies for Safety, Quality, and Convenience*, 125–150. Cham: Springer International Publishing.

Estonia – Ministry of Social Affairs. (2010). *Tervisekaitsenõuded toitlustamisele tervishoiu- ja hoolekandeasutuses* [Health Protection Requirements for Catering in Healthcare and Welfare Institutions]. Retrieved from www.riigiteataja.ee/akt/13251889

Filppa, M., & Kotilainen, M. (2012). *Kylmävalmistus ruoantuotantomenetelmänä – Esimerkkinä Tuusulan kunnan kylmävalmistusprosessin kehittäminen* [Cook Cold as a Food Preparation Method. Developing the Cook Cold Process in Tuusula Municipality]. Laurea University Degree Thesis. Retrieve from www.theseus.fi/bitstream/handle/10024/48915/FilppaKotilainen31lokakuuta2012.pdf?sequence=1&isAllowed=y

Global Survey of School Meal Programs. (2021). *Republic of Poland School Meal Programs*. Retrieved from https://gcnf.org/wp-content/uploads/2022/09/Poland_2021_final1.6.pdf

Hamann, K. (2021, 15 November). *StratKIT Sustainability Toolkit for Greener Public Procurement and Catering Services in Baltic Sea Region and Beyond*. Retrieved from www.sustainable-public-meal.eu/documents/13/StratKIT_Sustainability_Toolkit_report.pd

Haridus. (2022, 14 November). *Ligi 70 kooli Eestis asub võitlema toiduraiskamise vastu* [Nearly 70 Schools in Estonia are Starting to Fight Against Food Waste]. Retrieved from https://haridus.postimees.ee/7647557/ligi-70-kooli-eestis-asub-voitlema-toiduraiskamise-vastu

Holmbeck, P. (2020, December). *Best Practice in Organic Public Procurement: The Case of Denmark*. Retrieved from https://paulholmbeck.com/wp-content/uploads/2021/06/IFOAMOE_Best-Practice-in-Organic-Public-Procurement_The-case-of-Denmark.pdf

Landbrug, & Fødevarer. (2017, 2 August). *Markedsanalyse: Færre madpakker – men fortsat populært* [Market Analysis: Fewer Lunches – but Still Popular]. Retrieved 15 August, 2023, from www.ernaeringsfokus.dk/media/rwkku5ri/2017-faerre-madpakker-men-fortsat-populaert.pdf

Lee-Ross, D. (1999). A comparative survey of job characteristics among chefs using large and small-scale hospital catering systems in the UK. *Journal of Management Development*, *18*(4), 342–350. doi: 10.1108/02621719910265531

Lintukangas, S., & Palojoki, P. (2016). *School Dining in Finland*. Porvoo: Bookwell.

Luomuravintola. (2023). *Portaat luomuun -ohjelma* [Steps to Organic Program]. Retrieved from www.luomuravintola.fi/ohjelman_toiminta

Madkulturen. (2022). *Et stærkt samfund kræver en stærk madkultur i skolen* [A Strong Society Requires a Strong Food Culture in School]. Retrieved 22 March, 2023, from www.madkulturen.dk/madtanken/

Maietta, O. W., & Gorgitano, M. T. (2016). School meals and pupil satisfaction. Evidence from Italian primary schools. *Food Policy*, 62, 41–55. doi: 10.1016/j.foodpol.2016.04.006

Mikkelsen, B. E., & Sylvest, J. (2012). Organic foods on the public plate: technical challenge or organizational change? *Journal of Foodservice Business Research*, *15*(1), 64–83. doi: 10.1080/15378020.2011.650541

Mikkola, M., & Ala-Karvia, U. (2020, 30 December). *Baltic Sea Region Dynamic Sustainability Model for Public Procurement and Catering Services O3.4 (WP3)*. Retrieved from www.stratkit.eu/documents/34/BSR_Dynamic_Sustainability_Model_for_public_meals_final.pdf

Nuutila, J., Risku-Norja, H., & Arolaakso, A. (2018). Public kitchen menu substitutions increase organic share and school meal sustainability at equal cost. *Organic Agriculture*, 1–10. doi: 10.1007/s13165-018-0215-0

Ogrodinfo. (2019, 3 September). *Żywność ekologiczna w szkołach w Tarnogrodzie* [Organic Food in Schools in Tarnogród]. Retrieved from www.ogrodinfo.pl/sadownictwo/zywnosc-ekologiczna-w-szkolach-w-tarnogrodzie/

Olesen, A. H. (2017, 15 March). *Økologi er blevet billigere* [Organic Has Become Cheaper]. Retrieved 20 March, 2017, www.madforlivet.com/sund-inspiration/oekologi-blevet-billigere/

Poland – Health Ministry. (2016). *Rozporządzenie Ministra Zdrowia z dnia 26 lipca 2016 r. w sprawie grup środków spożywczych przeznaczonych do sprzedaży dzieciom i młodzieży w jednostkach systemu oświaty oraz wymagań, jakie muszą spełniać środki spożywcze stosowane w ramach żywienia zbiorowego dzieci i młodzieży w tych jednostkach* [Regulation of the Minister of Health of 26 July 2016 on Groups of Foodstuffs Intended for Sale to Children and Youth in Units of the Education System and the Requirements to be Met by Foodstuffs Used as Part of Mass Catering for Children and Youth in These Units]. Retrieved from https://dziennikustaw.gov.pl/DU/rok/2016/pozycja/1154

Proveg. (2020, 12 March). *Międzynarodowy dzień posiłków szkolnych: jak wprowadzić więcej warzyw do stołówek* [International School Meal Day: How to Get More Vegetables into Canteens?]. Retrieved from https://proveg.com/pl/blog/miedzynarodowy-dzien-posilkow-szkolnych-jak-wprowadzic-wiecej-warzyw-do-stolowek/

Ruge, D. (2021, 5 January). *Varmt madtilbud på alle skoler i Århus Kommune* [Hot Food Offerings at All Schools in Aarhus Municipality]. Retrieved from www.folkeskolen.dk/1862230/varmt-madtilbud-paa-alle-skoler-i-aarhus-kommune

Sønderborg. (2019). *Handleplan for Fødevarer 2019–2021 Sønderborg Kommune som klimavenligt fødevareområde* [Food Action Plan 2019–2021, Sønderborg Municipality as a Climate-Friendly Food Area]. Retrieved from https://sonderborgkommune.dk/sites/all/files/Forvaltninger/Intern%20Stab/Beredskab/Dokumenter/handleplan_for_foedevarer_2019-2021.pdf

StratKIT. (2020, 25 September). *Innovative Strategies for Public Catering: Sustainability Toolkit Across BalticSea Region (Work Package 2, Output 2.4)*. Retrieved from www.stratkit.eu/documents/4/Joint_BSR_Report_for_Sustainable_PPCS.pdf

Taimse Teisipäeva. (2021, 15 February). *30 kooli üle Eesti on liitunud taimse koolilõuna programmiga*. [30 Schools Across Estonia Have Joined the Plant-Based School Lunch Program]. Retrieved from https://taimneteisipaev.ee/uudised/30-kooli-ule-eesti-on-liitunud-taimse-koolilouna-programmiga

Tikkanen, I., & Urho, U. M. (2009). Free school meals, the plate model and food choices in Finland. *British Food Journal*, *111*(2), 102–119. doi: 10.1108/00070700910931940

Tyczyńska, K. (2016). *Ekologiczna żywność w państwowych placówkach oświatowych – jednak jest możliwa* [Organic Food in State Educational Institutions – It Is Possible]. Retrieved from https://www.zdrowyprzedszkolak.org/ekologiczna-zywnosc-w-panstwowych-placowkach-oswiatowych-jednak-jest-mozliwa/

Vetemaa, A., Mikk, M., & Peetsmann, E. (2022). *Mahepõllumajandus eestis Organic farming in estonia 2021*. Estonian Organic Farming Foundation. Retrieved from www.mahekubi.ee/upload/Editor/mahe_eestis_2021_est_eng_digi.pdf

Wasak, M. (2023). *Zdrowa żywność wprost od rolników znów może pojawić się w szkołach – video* [Healthy Food Directly from Farmers May Appear in Schools Again – Video]. Retrieved from www.wrp.pl/zdrowa-zywnosc-wprost-od-rolnikow-znow-moze-pojawic-sie-w-szkolach-video/

5 Sustainable public food procurement in France

Isabelle Hasquenoph

Introduction

For some years now, public policies implemented in France have been giving increasing attention to the issue of sustainable food. Since 2018, several laws have been adopted to ensure sustainable and quality food,[1] fight against food waste,[2] ensure a fair remuneration to farmers,[3] and transparency of the origin of agricultural and food products.[4] This is a current concern which is of interest to France and many countries in the world. These public policies align with the European ambition to make food system "fair, healthy and environmentally friendly" and to improve public procurement's sustainability (European Commission, 2019, 2020, 2022).

These public policies impact public procurement law, which is now used to promote sustainable, high-quality food. Public procurement law has a driving role to play in shifting our food system towards a system that is more respectful of health and the environment.

This trend is part of the more general framework of the instrumentalisation of public procurement. It allows questioning the evolution of the functions of public procurement law over the centuries (Arrowsmith et al., 2011). Public procurement law in France dates back to the Middle Ages. In the 14th century, the purpose of procurement procedures was to protect public money and public morality. A royal decree of 4 July 1376 stated that their objective was to avoid "fraud and malice. . . to the detriment of the King and the public good" (Bezançon, 1999, p. 195). These functions have been reaffirmed in history, until the 21st century which constitutes a turning point. The reform of public procurement law initiated by the European Union in 2014 bears witness to the changes in the functions of public procurement law. This reform led to the adoption of the French *Code de la commande publique* (Public Contracts Code – CPP), which came into force on 1 April 2019. Today,[5] public procurement law serves "horizontal policy goals":[6] social and environmental objectives,[7] a social and solidarity economy,[8] the transition to a circular economy,[9] and sustainable food.[10] The Climate and Resilience Act of 22 August 2021 clearly displays this change in the objectives of public procurement. It added a new article 3–1 to the CPP, which states that "Public procurement participates in achieving the objectives of sustainable development, in their economic, social and environmental dimensions". This is a step forward compared to the European

DOI: 10.4324/9781003393023-7

directives: despite doctrinal proposals (Andhov et al., 2020) achieving sustainability targets has not been added to the aims of the Public Procurement Directives, which only refers to the achievement of the internal market. Now the objectives of sustainable development are set at the same level as the objective of opening up the national procurement market to competition (Richer, 2015). However, the question of the effectiveness of implementing these objectives remains.

This topic is also relevant for citizens: consumers are increasingly sensitive to the conditions in which their food is produced. They want to consume better quality, ecological and healthy products. School canteens are often the focus of concern, due to the sensitivity of the public concerned. In France, the first school canteen was created in 1844, under the July Monarchy, in the city of Lannion (Nourrisson, 2004). Local initiatives multiplied thereafter: in 1860 in Pont-Audemer, the Bon Secours nuns "prepared hot food that they brought to the children of the three schools of the city" (Gosselin, 1908). In 1870, the revolutionary activist Louise Michel created a canteen in the school she founded in Montmartre. The development of school canteens was given new impetus when Jules Ferry made school compulsory in the 1880s (Robin, 2017, p. 82). However, the conditions of hygiene and nutrition were deplorable: the refectories were covered courtyards, the menus consisted mainly of bread, soup and mashed vegetables,[11] and the children could drink wine or cider. From the 1920s to the 1930s, and especially after World War II, the school provided a remedy for children's nutritional deficiencies (for example, by distributing a glass of milk to children up to 12) (Parayre, 2017). At the end of the 1950s, 800,000 children were attending school canteens, and a real improvement could be observed in the menu composition.[12] In 1956, the consumption of wine was prohibited in school canteens (Nourrisson, 2017).

It is worth noting that the school catering service is closely linked to the decentralised organisation of France. The *communes* (municipalities) manage nursery and primary schools (there are 34,965 *communes* in France). In these schools, the *communes* are free to organise or not a school catering service: school catering is an optional public service at the communal level. However, for the departments responsible for secondary schools ("*collèges*"), and the regions responsible for high schools ("lycées"), school catering is a compulsory public service (Art. L. 213-2 and L. 214-6 C. Education).

Sustainable food procurement in France is thus of renewed interest. French jurists have written little on this topic. From a methodological point of view, sustainable food procurement allows exploring the borders between law and other disciplines such as economy or geography. Firstly, a 2017 economics thesis explores the "Make or Buy" decisions in the catering sector and the relationship between public food procurement contracts and political contestation (Robin, 2017). Secondly, sustainable food is also of interest to geographers. The "Law and Geography" movement is less developed than the "Law and Economics" movement, but it is promising, especially in public procurement law (Blomley, 1994; Hasquenoph, 2021). This movement examine the mutual influences between law and geography. Indeed, "a change in the geographical environment in itself might change the function of a given legal institution", and law creates and modifies space (Grossfeld,

1984; Forest, 2009, p. 28). Integrating this geographical approach into the study of sustainable food procurement makes it possible to develop a more global vision of the issues at stake. Geographers are interested, for example, in short supply chains, which they consider to be "territorial innovations" creating new local dynamics (Chiffoleau & Prévost, 2012). They study the influence of the environment on obesity and propose ways to reform (Pech, 2021).

Specialists in public procurement law can take advantage of these interdisciplinary approaches to enrich their research (Champeil-Desplats, 2014). The field of sustainable development and food is particularly well suited to these approaches.

This study analyses the development of sustainable public food procurement in France. The first section examines the impact of public policies promoting sustainable food on public procurement law. The second section addresses the adequacy of public procurement law to the challenges of sustainable food.

The impact of public policies promoting sustainable food on public procurement law

Since 2014, the National Food Programme has aimed to encourage the supply of seasonal or organic agricultural products to public and private catering (Clamour, 2018). Subsequently, the 2018 Egalim law marked a turning point in promoting sustainable food in collective public catering. It has since been supplemented by the AGEC law on the circular economy and the fight against waste, and the Climate and Resilience Law which aims to support the transformation of the French food and agricultural model. These different laws have impacted public procurement law. This section presents the measures promoting sustainable food purchasing and analyses their implementation.

Measures promoting sustainable food procurement in France

The public authorities' will to promote sustainable food has led to a proliferation of law enactments. The public purchaser will only be able to meet the new requirements by a combined reading of the CPP and the Egalim, AGEC, Climate and Resilience laws, the provisions of which have been codified in the CPP but also in the *Code de l'environnement* (Environment Code) or the *Code rural et de la pêche maritime* (Rural and Maritime Fishing Code, henceforth Code rural). This superposition of texts breaks with the unity sought by the legislator when drafting a CPP. Moreover, it may complicate the task of the public purchaser. These new provisions lead to the emergence of a public procurement law "outside" this Code, which the public purchaser must not ignore (Vaysse, 2021).

Promoting the purchase of quality food products

One of the emblematic measures of the Egalim law requires that by 1 January 2022, the meals served in collective public catering (collective restaurants run by entities governed by public law) shall include a share at least equal in value to 50% of quality and sustainable products, of which 20% (in value) shall be organic products (art. L. 230–5–1-I Code rural). For beef, pork, sheep, poultry and fishery products, the significant share of these quality products is raised to 60% by 1 January 2024 at the latest. This rate is 100% in collective restaurants managed by the State (art. L. 230–5–1 I Code rural). These requirements must be taken into account by public purchasers when defining their needs.

These quality and sustainable products are for example: products acquired in a way that takes into account the costs of the environmental externalities associated with the product during its life cycle; products whose procurement is based primarily on environmental performance and the development of direct procurement of agricultural products; Fairtrade products; products with quality labels (the red label, the "farm product" label, the eco-label "sustainable fishing") or products "satisfying. . . in an equivalent manner, the requirements defined by these signs, eco-labels or certification". Two elements are worth noting. Firstly, the phrase "or equivalent" gives flexibility to the public purchaser but remains vague. Secondly, local products are not explicitly mentioned in this list.

These new obligations come with a certain degree of transparency. For example, the law sets out a duty to inform users of the proportion of sustainable quality products in the composition of meals, by display or electronic means (art. L. 230-5-3 Code rural). In addition, the Government must submit to Parliament and make public a statistical report on the application of article L. 230-5-1 of the Code rural. This requirement for transparency is also reflected in the new measures aimed at ensuring the traceability of meat origin. Under a decree of 26 January 2022, the origin of meat served in out-of-home catering must be indicated from 1 March 2022. This is an experiment until 29 February 2024, which is of direct interest to public purchasers.

Promoting low greenhouse gas-emitting food

The legislator's will to promote food whose production emits little greenhouse gas has resulted in generalising the obligation to offer a vegetarian menu in collective catering. Indeed, reducing meat consumption contributes to the fight against global warming (Godfray et al., 2018). Thus, public or private managers of school catering services must now offer at least once a week a vegetarian menu (art. L. 230-5-6 I Code Rural). On an experimental basis, voluntary local authorities may offer a daily vegetarian menu, while ensuring the nutritional balance of the meals and respecting requirements adapted to the needs of users. The results of this experiment will be presented by the Government to Parliament (art. L. 230-5-6 II Code Rural). Finally, by 1 January 2023 at the latest, managers of collective catering services of the State, must offer a daily vegetarian menu, when they usually offer a multiple choice of menus (art. L. 230-5-6 III Code Rural).

The fight against climate change has also led to ban plastic food containers in public catering (art. L. 541-15-10 Environment Code). By 1 January 2020 at the latest, the law required the ending of the use of plastic water bottles in school canteens, and by 1 January 2025 at the latest, the end of the use of plastic food containers for cooking, heating and serving in school and university canteens and establishments catering for children under six years of age. The deadline is 1 January 2028 for local authorities with fewer than 28,000 inhabitants.

Limiting food waste

At the same time, France has adopted various measures to combat food waste in collective catering, particularly public catering. Food waste has been defined as "Any food intended for human consumption which, at any stage of the food chain, is lost, thrown away or degraded" (art. L. 541-15-4 Environment Code). The AGEC Law has set a national goal of reducing food waste by half by 2025, compared to its 2015 level in food distribution and collective catering (Ministry of Ecological Transition and Territorial Cohesion, 2023). Figures on food waste are alarming (Ecological Transition Agency, 2020). In France ten million tons of consumable food are wasted yearly, which is equivalent to 150 kilos per year per person. In collective catering, the average waste is 120 g per person per meal. The figures vary according to the type of institution involved: hospitals have the highest wastage (170 g/person/meal), followed by retirement homes (155 g/person/meal) and school canteens (middle schools and high schools), with 120 g/person/meal. Companies are more efficient (90 g/person/meal). Waste does not only occur at the consumer level. It also occurs in the kitchen (overestimation of quantities, poor stock management) and during service (unsuitable portions served).

In 2015,[13] an obligation to fight against food waste was imposed on public entities managing collective catering services. In 2019[14] this obligation was extended to all private collective catering operators, and two obligations were added. Firstly, set up measures to diagnose wastage (what are the quantities lost, the components affected, are the grammages adapted etc). Secondly, operators preparing more than 3,000 meals a day must propose a convention to charities to donate unsold consumables.

The Climate and Resilience Law also includes measures to combat food waste. On an experimental basis for three years and on a voluntary basis, the law allows public catering to introduce an online booking system to adapt the supply to the number of meals actually needed (art. 256).[15] A regulation must define the reservation deadlines, reservation period, reservation type (type of menu, menu with a main course and dessert, allergies), reservation modalities (confirmation of presence on the same day, automatic reservation for the year, reservation on the internet, confirmation by phone. . .), payment modalities.[16]

In addition, new objectives have been added to the National Food Programme, which "takes into account food sovereignty, social justice, food education for young people, in particular the promotion of food-related know-how and the fight against food waste" (art. L1 III Code Rural).

Finally, it can be noted that the COVID-19 crisis led to the adoption of emergency measures to combat food waste. A decree of 22 July 2020 temporarily raised the threshold below which public food procurement contracts could be awarded by direct contracting to 100,000 euros. The objective was to allow the disposal of food products stored during the lockdown of spring 2020.

There is a real ambition of the public authorities to promote sustainable food, and to make public procurement a tool to achieve this goal. However, a discrepancy remains between the ambition stated and the implementation of the measures.

Limits: The difficulties of implementing these measures

The implementation of these measures raises practical and legal difficulties.

Practical difficulties

The emblematic measure of the Egalim law, which requires that meals served in public catering include at least 50% quality products, including 20% organic products, is not uniformly respected. Indeed, there are major disparities between cities and between establishments.

Researchers have proposed to transpose the AMO (ability, motivation and opportunity) model by Boxall and Purcell to examine public buyers' ability, motivation and opportunity to develop sustainable food procurement practices (Boxall & Purcell, 2011; Andhov et al., 2020). They show that commitment to change can occur if public purchasers are required to implement sustainable public procurement in their work, due to strict rules. But for the time being no sanctions are attached to disregarding the obligations laid down by the texts. Moreover, the objectives regarding sustainable food purchases are often included in soft law tools (plans, sustainable food schemes[17]), which are not legally binding (Delzangles, 2023).

What is more, this measure has faced multiple contestations. The objective set has been seen to be too high. In 2019, the share of organic products in collective catering was only around 4.5% (National Observatory, 2021, p. 6). The implementation timeframe has been criticised as being too short (November 2021–1 January 2022), not leaving enough time for public purchasers to acquire the necessary skills. Finally, some reckon that this measure increases the cost of the purchase and thus put a strain on public finances.

These criticisms can be countered in several ways.

Firstly, the objective of 50% quality products, including 20% organic products, is not so high compared to public policies implemented in other countries (in Copenhagen, the objective of 90% organic products was reached in 2016). Secondly, on a sample of 6,766 canteens, a study showed that the share of organic products served was already 37%, *i.e.* 17% more than what the Egalim law imposes and much more than the 4.5% share in collective catering surveyed in 2019 (National Observatory, 2021). In some cities such as Paris, the target has been largely exceeded, with 53% of products from organic and sustainable sources in school canteens (Paris, 2021). Thirdly, a recent report[18] has shown that the obligations introduced by the Egalim law led to an additional cost of about 0.14 to 0.42 euros per meal. But this increase can be compensated by measures to fight against food waste or the introduction of vegetarian meals. In addition, even if some sustainable initiatives are more expensive, they might also be more cost-effective in the long run.

The introduction of vegetarian menus has also been contested, particularly in the city of Lyon. Supporters of this measure invoked the carbon footprint reduction, while opponents pointed out the risks for children from disadvantaged backgrounds and the agricultural sector. This example shows that, even if "the (top) management of contracting authority, including political elected head. . . plays a crucial role in creating ability, opportunity and motivation in the organizations for sustainable public procurement" (Andhov et al., 2020), they sometimes face political contestation.[19]

Finally, the conclusions of the National Sustainable Procurement Plan 2022–2025 (p. 11) are not optimistic. It notes that some purchasers, due to a lack of knowledge or a lack of means, do not dare to engage in this sustainable purchasing approach and limit themselves to repeating the wording of previous contracts.

In addition to these practical difficulties, legal difficulties remain.

Legal difficulties

The first difficulty is a question of timing (Vaysse, 2021). How can the new obligations be integrated into public food contracts with a deadline beyond 1 January 2022? This raises the issue of modifying contracts during their execution (Hoepffner, 2016, p. 280). Several instruments can be used to modify a public contract and deal with contingencies, be they legal, technical or economic (the review or option clause – art. R. 2194-1 CCP; the unilateral modification power of the public purchaser when the contract is an administrative one – art. L. 6 CCP).

The question of modifying contracts during performance has recently arisen again in connection with the increase in raw material prices following the health crisis and the war in Ukraine. A ministerial circular details the possibilities public procurement law offers to modify the financial conditions and duration of contracts in order to deal with "unforeseeable circumstances" (Economy Ministry, 2022).

The second issue is the conciliation between these objectives with the principles of public procurement. According to some authors, the reference to short supply chains, to costs imputed to environmental externalities linked to the product during its life cycle or to direct supplies of agricultural products (art. L. 230-5-1 Code Rural) raises questions about the principles of transparency and equal treatment and "even more so concerning the very effectiveness of the principle of competitive tendering" (Roux, 2019; Vaysse, 2021). But these concerns must be tempered. Public procurement law only prohibits the use of selection criteria related to the geographical location of operators.[20] And it offers tools that facilitate access to public procurement for local operators, without using such criteria. Their access to public procurement can positively impact the local economy. (Richer & Cantillon, 2017). Indeed, authors have shown that a regular and predictable public procurement allows small farmers to invest, increase and diversify their production (Swensson & Tartanac, 2020; Kelly & Swensson, 2017). It can be assumed that these positive repercussions on the local territory will in turn improve public purchasers' motivation and commitment to sustainable public food procurement. What is more, the new article 3-1 of the CPP shows that the objectives of sustainable development are as legitimate as the objective of opening up the national procurement market to competition (Richer, 2015).

The adequacy of public procurement law to the challenges of sustainable food

The CPP offers tools to meet these objectives and promote sustainable public food procurement development. But it implies a change of habits and the development of knowledge and skills, which have been identified "as essential drivers or conditions for the implementation of sustainable public procurement" (Andhov et al., 2020). To help public entities change their purchasing practices, the Localim "toolbox" (guidance document) of the Ministry of Agriculture, offers public purchasers concrete examples for development of local and quality food procurement (Localim, 2016).

Provisions of the *Code* enable the purchaser to buy sustainable food at all stages of the procedure.

The preparation phase: Defining needs

"Sourcing" [early market engagement]

To properly define the needs for purchasing sustainable food products, the public buyer must have a good knowledge of the local economic market and what is available from suppliers. This concept is referred to in French as "sourcing". This is encouraged by certain legal provisions – art. R. 2111-1 CCP; art. 40 dir. 2014/24) (Dyens, 2016). In France, the Association of Mayors of France has adopted a "charter for the promotion of culinary and agricultural heritage" which seeks to promote dialogue between all the actors in the chain, to develop the local offer of products from organic farming (French Mayors' Association, 2016). The city of Lons-Le-Saunier, for example, works in conjunction with the local organic sector, which allows the supply to be structured according to demand (Perrier, 2017). In the city of Lorient, the municipality identified and visited suppliers' sites to prepare calls for tenders "as close as possible to their constraints and possibilities" (Bréger, 2019).

The "sourcing" technique is well suited for an analysis from a "law and geography" perspective: facilitating access for local producers to public procurement can positively impact on local economy and lead to a recomposition of space. Besides, this technique shares similarities with the methods used in geography (collect data and draw up a "cartography" of the offers available within a certain perimeter).

Labels

The public buyer can refer to labels to buy quality products. He may refer to them in particular to meet the requirement laid down by the Egalim law to serve meals including a share at least equal in value to 50% of quality products, including 20% of products from organic farming.

Indeed public purchasers may require a particular label in the technical specifications, award criteria or performance conditions. The label used must meet certain characteristics (R. 2111-14 CCP) and have a link with the subject matter of the contract (R. 2111-15 CCP).

These conditions are necessarily met for the labels set out in the Egalim law (Clamour, 2018). To obtain supplies of quality products, the buyer may refer to the "organic farming" labels (Regulation 832/2007 EC), or quality labels: Red Label, Designation of Origin or Geographical Indications. Labels can also be used to develop the use of Fairtrade products.

In addition, the public buyer must accept any "equivalent", *i.e.* labels other than those he has referred to but which guarantee that the products meet equivalent requirements (art. R. 2111-16 CCP). Producers may refer to more demanding labels (Clamour, 2018). For example, the trademark "Bio cohérence" requires 100% organic ingredients (compared to 95% for the European label).

CHOICE OF CONTRACTUAL MODEL

Upstream: The "make or buy" decision

Public entities always have a choice between contracting out services or providing those themselves.[21]

In the field of collective catering, and particularly school canteens, the question of how to manage the service arises. Several cities are considering taking over the school catering service or entrusting the management of the service to a *Société publique locale*, a publicly owned company created by a Law of 28 May 2010 to meet the in-house contract conditions.[22]

Direct provision is seen to have multiple benefits for the management of school canteens. Indeed, public contracts are necessarily concluded for a limited period of time. As a result, public purchasers might have a short-term vision that can conflict with the long-term objectives of public policies to combat climate change and promote sustainable food systems (Vaysse, 2021). Direct management makes it possible to integrate this long-term vision. It should be noted, however, that the municipality will necessarily have to conclude contracts, at least for food procurement. It is indeed unlikely that it will manage the whole chain, from the vegetable garden to the plate.

In his economics thesis on school canteen procurement, Y. Robin sheds light on the determinants of the choice between direct provision and externalisation. He assumes that in-house provision is "more transparent toward quality". Thus, the municipality might choose a direct management when citizens are sensitive to quality. But he also demonstrates that the choice between different modes of service provision depends on political contestation. When the municipal majority is strong, it is less exposed to political contestation about probity or mismanagement of public funds. Thus, it has more incentive to internalise the service. On the contrary, choosing outsourcing can be a way to signal "probity" and "limit speculation on the discretionary use of money" (Robin, 2017).

If the public entity chooses to use contracts to meet its needs, it can enter into two types of contracts.

Public procurement or concession

Firstly, the public entity can use public procurement: a public supply contract (for example for the supply of fruits and vegetables) or a public service contract (for example for assistance in the production of meals on the beneficiary's site or the preparation and delivery of meals).

Secondly, the public entity may also resort to a public service concession if its co-contractor bears an operating risk. The problem is that the administrative judge assesses the exploitation risk casuistically. In a 2017 ruling,[23] the French requalified the management contract for a public catering service as a public procurement contract instead of a concession since the economic operator was not considered to bear a risk because the number of users was not likely to vary substantially during the execution of the contract.

Division of the contract into lots

Allotment can facilitate market access for small suppliers and "promote the purchase of quality and environmentally friendly products" (Localim, 2016, p. 2). The allotment can help develop the local organic industry and contribute to the dynamism of the regional economy. For example, the city of Lorient has divided contracts into many lots (32 lots for contracts between 2014 and 2017). Several lots were organic, reserved for short supply chain or labelled products (Bréger, 2019).

Contract award procedure

Design of the call for competition

The Localim Guide recommends favouring procedures "with negotiation" for public food procurement, as suppliers "may be small structures not accustomed to formalism".

Among the "formalied" procedures, the procedure with negotiation can only be used in six cases (art. R. 2124-3 CCP). For example, when the need cannot be met without adapting immediately available solutions (1°); or when the need consists of an innovative solution (2°).

In addition to the formalised procedures, the Code includes other possibilities for negotiation.

Firstly, the adapted procedure (art. R. 2123-1 CCP) may be used for public contracts below the thresholds, and for contracts whatever their value, relating to "social services and other specific services" (and in particular collective catering services: art. R. 2131-7; Appendix 3 CCP). Secondly, purchasers can also resort to the negotiated procedure without publicity or competition, but this only concerns specific cases (art. R. 2122-1 to R. 2122-11 CCP). Some of them could be of interest to sustainable public food procurement. Indeed, this procedure can be followed for contracts with an estimated value of less than 40,000 or the purchase of innovative services (under 100,000 euros).

Award criteria

The combination of certain award criteria might also facilitate the access of local suppliers to public food procurement.

The public contract must be awarded to the operator who submits the "most economically advantageous tender" (art. L. 2152-7 CCP; art. 67 dir. 2014/24). This offer is assessed on the basis of one or more criteria. The Climate and Resilience Law introduced an innovation: "at least one of these criteria shall take into account the environmental characteristics of the tender" (art. L. 2152–7 CCP). From 2026 onwards, the public purchaser will no longer be allowed to refer to the sole criterion of price: if he chooses to use only one selection criterion, it can only be the criterion of life-cycle costing integrating environmental concerns.

The public buyer can assess the offer with regard to the life cycle costing" (art. R. 2152-7 1° CCP; art. 68 dir. 2014/24) (Dragos & Neamtu, 2013). This "life-cycle costing" covers a variety of costs, including the cost imputed to environmental externalities, including the cost of emissions of greenhouse gases and other pollutant emissions and other climate change mitigation costs. This concept is sometimes considered to be difficult to implement by public purchasers.

It can be interesting to rely on a plurality of award criteria. For instance, quality, diversity, production and marketing conditions, environmental protection performance, biodiversity, animal welfare, developing direct supplies of agricultural products (art. R. 2152-7 CCP). The objective of this latter criterion is to create "short supply chains". "Short" does not mean "local": it designates the direct sale from the producer to the consumer, or the indirect sale but on the condition that there is only one intermediary. The notion of short supply chain is approached through a relational and not spatial angle (Cantillon, 2022). Therefore, the use of this concept does not, in principle, favour a local operator. But in practice, the requirement for organic, homemade or seasonal products can lead cities to source locally, which directly impact land organisation, and employment, and encourages local suppliers to convert to organic farming. For example, in the 14th arrondissement of Paris, five criteria are implemented to ensure the supply of quality and sustainable food in school canteens. Price represents 35% of the total score, quality 25%, respect for grammages 20%, respect for packaging 15% and deadlines 5% (Robin, 2017, p. 24). The quality criterion offers flexibility to public purchasers and may allow them to favour a local supplier over an international one (Robin, 2017).

In Brittany, sustainable food procurement has been used to protect water resources. Two public entities have launched a public procurement contract for the supply of food products, aiming to "promote local and efficient agriculture" in order to protect the water quality of the basins that supply the city of Rennes.

The criteria of sustainability represented 50% of the total score, the criterion of quality of the products 30% and the price 20% (Reseco, 2019).

In general, it should be noted that France has adopted measures to promote short supply chains for several years (Maréchal, 2008; Cantillon, 2022; Villeneuve, 2018).

The National Food Programme indeed encourages "short supply chains and geographical proximity between agricultural producers, processors and consumers" (art. L1 Code rural). Since 2014,[24] the territorial food projects (art. L. 111-2-2 Code rural) have been a privileged tool for pursuing these objectives and implementing the food policy in the territories. These territorial food projects are developed voluntarily by the actors of a territory and aim to "relocalize agriculture and food in the territories by supporting the installation of farmers, short supply chains or local products in the canteens" (Agriculture Ministry, 2022). The Egalim law invites public legal entities in charge of collective catering to "develop the acquisition of products within the framework of territorial food projects" (art. L. 230-5 C. rural). By January 1, 2023, the Ministry of Agriculture recognised nearly 400 territorial food projects (Agriculture Ministry, 2023).

Variants can also be attractive because they can improve the adaptation of the offer to certain goals pursued, with proposals for more qualitative products or better environmental performance, which the buyer was unaware of (Localim, 2016).

Finally, the Localim guide (p. 5) recommends testing products, i.e. that purchasers require in the consultation documents that a sample is attached to the offer. In this case, an award criterion based on the quality of the sample must be provided. This has been the case in the 14th arrondissement of Paris.

Contract performance

The Climate and Resilience Law has modified the rules relating to the contract performance conditions. From now on, "the contract performance conditions shall take into account considerations relating to the environment" (art. L. 2112–2 CCP).

Control of contract performance is crucial, but its insufficiency is often criticised (Hoepffner, 2018). It is as if public buyers consider the procurement procedure sufficient to ensure the proper execution of the contract. The control of the contract's performance may relate to the nutritional balance, the organoleptic quality of the dishes, the origin and quality of the products and traceability (Localim, 2016). Failure by the service provider to comply with its obligations may be subject to the sanctions provided for in the contract, such as penalties. Apart from sanctions, the Localim Guide also stresses the importance of dialogue with the various stakeholders (technical teams, guests) with a view to continuously improve services.

Conclusions

To conclude, there is a real political ambition to promote sustainable food in France, and to make public procurement law a tool to achieve this goal. Even if discrepancies sometimes remain between the measures adopted and their implementation,

French law now offers tools for sustainable food public procurement. Public buyers should be encouraged to use them.

Sustainable public food procurement has been largely ignored in theoretical studies in recent years in France, but there is no doubt that it will lead to the development of new research. Indeed, this topic is well suited for multidisciplinary studies: it lies at the intersection of law and politics, economics and geography. It also relates to important issues such as climate change, economic development, human health and democracy.

Notes

1 *Egalim* Law n°2018–938, October 30th, 2018; *Climat et Résilience* Law n°2021-1104, August 22nd, 2021.
2 *AGEC* Law n°2020-105, February 10th, 2020.
3 *Egalim* Law; *Egalim 2* Law n°2021-1356, October 18th, 2021.
4 Law n°2020–699, June 10th 2020; Decree n°2022-65, January 26th, 2022.
5 Before, the ECJ authorised public purchasers to take social and environmental consideration into account in public procurement procedure in ECJ, September 20th, 1988, *Beentjes*, C- 31/87; ECJ, September 17th, 2002 *Concordia Bus Finland*, C-513/99.
6 Arrowsmith (2010).
7 Llorens & Zimmer (2014).
8 Clamour (2014).
9 Art. L. 541-1 C. Environnement.
10 *Egalim* Law; Clamour (2018).
11 Nourrisson, D. (2020). However, a school canteen regulation from the 12th arrondissement of Paris from 1898 shows that slightly more diversified menus could also be offered to students: sausages with vegetables on Thursday, potatoes with bacon on Friday, for ten cents per portion (Adenis, 1898).
12 Parayre, S. (2017). In 1957 in the 13th arrondissement of Paris, the menu proposed included asparagus in vinegar as a starter, a veal escalope and a gratin of coquillette for the main course, followed by sweet cottage cheese and cherries.
13 Law n°2015-992, August 17th, 2015.
14 Ord. n°2019-1069, October 21st, 2019.
15 Decree n°2022-840, April 4th, 2022, *relatif à l'expérimentation de solutions de réservation de repas en restauration collective*. [on the experimentation of meal reservation solutions in collective catering]. For implementation guidance see Ma-Cantine (2021).
16 The prices of meals served in school canteens are set by local authorities, often considering the family income. Faced with inflation, some municipalities have taken steps to ensure that meals served in primary schools are free of charge: this is the case for example in Saint-Denis (www.lesechos.fr/pme-regions/ile-de-france/malgre-linflation-saint-denis-etend-la-gratuite-des-cantines-aux-classes-de-cp-1785471). The issue of free school canteens is a current topic in France: a bill was submitted in September 2022 to introduce free school canteen meals in middle and high schools (www.assemblee-nationale.fr/dyn/16/textes/l16b0240_proposition-loi#).
17 Plan national pour des achats durables (PNAD 2022–2025); Schéma de promotion des achats publics socialement et économiquement responsables (SPASER – art. L. 2111-3 CCP).
18 Information Report on behalf of the Economic Affairs Committee, n°89, October 2019.
19 More recently, farmers' unions challenged the decision of the mayor of Lyon to introduce a single meat-free menu in school catering services for a period of five weeks to consider health constraints linked to the COVID crisis. In a judgment of 23 March 2023, the Lyon Administrative Tribunal ruled in their favour and annulled the mayor's decision (TA Lyon, March 23rd, 2023, n°2101388–21001390, *FDSEA du Rhône*; JCPA n°15, 17 April 2023, 2128.

20 CE, July 29th, 1994, Commune de Ventenac-en-Minervois, n°131562; ECJ, Oct. 27th, 2005, Commission c/Spain, C-158/03.
21 Dir. 2014/24, recital 5; art. L. 1 CCP; CJEU, October 30th, 2019, *Irgita*, C-285/18.
22 ECJ, November 18th, 1999, *Teckal srl*, C-107/98.
23 CE, May 24th, 2017, *Société Regal des îles*, n°407213.
24 Law n°2014–1170, October 13th, 2014, *Loi d'avenir pour l'alimentation, l'agriculture et la forêt*. [Law on the future of food, agriculture and forestry].

Bibliography

Laws & court rulings

A number of French laws and court rulings are referred to in the end notes. Also European Union Directives and rulings of the European Court of Justice.

Government publications

Agriculture Ministry. (2022, 5 May). *Qu'est-ce qu'un projet alimentaire territorial*? [What is a territorial food project?]. Retrieved from https://agriculture.gouv.fr/quest-ce-quun-projet-alimentaire-territorial

Agriculture Ministry. (2023). *Près de 430 Projets Alimentaires Territoriaux (PAT) reconnus par le ministère au 1er avril 2023*. [Nearly 430 Territorial Food Projects (PAT) recognized by the ministry as of April 1, 2023]. Retrieved from https://agriculture.gouv.fr/pres-de-400-projets-alimentaires-territoriaux-pat-reconnus-par-le-ministere-au-1er-janvier-2023

Ecological Transition Agency. (2020). *La France est engagée à réduire de moitié le gaspillage alimentaire d'ici 2025*. [France is committed to halving food waste by 2025]. Retrieved from www.ademe.fr/expertises/dechets/passer-a-laction/eviter-production-dechets/dossier/reduire-gaspillage-alimentaire/enjeux

Economy Ministry. (2022, 4 December). *Prise en compte de l'évolution des prix des denrées alimentaires dans les marchés publics de restauration* [Taking into account the evolution of food prices in public catering markets]. Retrieved from www.economie.gouv.fr/daj/prise-en-compte-de-levolution-des-prix-des-denrees-alimentaires-dans-les-marches-publics-de-0

Localim. (2016, 24 October). *Localim: la boîte à outils des acheteurs publics de restauration collective* [Localim: the guide for public buyers of collective catering]. Retrieved from https://agriculture.gouv.fr/localim-la-boite-outils-des-acheteurs-publics-de-restauration-collective

Ma-Cantine. (2021). *Guide d'accompagnement – décret d'expérimentation de solution deréservation de repas en restauration collective* [Support guide – experimentation decree for meal reservation solutions in collective catering]. Retrieved from https://ma-cantine.beta.gouv.fr/static/documents/Guide_pratique_XP_RESERVATION.pdf

Ministry of Ecological Transition and Territorial Cohesion. (2023, 3 January). *Gaspillage alimentaire* [Food waste]. Retrieved from www.ecologie.gouv.fr/gaspillage-alimentaire

Other references

Adenis, C. (1898). *Les cantines scolaires du 12ème arrondissement* [School canteens in the 12th arrondissement].

Andhov, M., Caranta, R., Stoffel, T., Grandia, J., Janssen, W. A., Vornicu, R., . . . & Wiesbrock, A. (2020). *Sustainability through public procurement: the way forward – reform proposals*. Retrieved from https://papers.ssrn.com/sol3/papers.cfm?abstract_id=3559393

Arrowsmith, S. (2010). Horizontal policies in public procurement: a taxonomys, *Journal of Public Procurement*, 10(2), 149–186. doi: 10.1108/JOPP-10-02-2010-B001

Arrowsmith, S., Treumer, S., Fejo, J., & Jiang, L. (2011). *Public procurement regulation: an introduction*. EU Asia Inter University Network, 3. Retrieved from www.nottingham.ac.uk/pprg/documentsarchive/asialinkmaterials/publicprocurementregulationintroduction.pdf

Bezançon, X. (1999). *Essai sur les contrats de travaux et de services publics. Contribution à l'histoire administrative de la délégation de mission publique* [Essay on public works and services contracts. Contribution to the administrative history of delegation of public mission]. Paris: LGDJ.

Blomley, N. K. (1994). *Law, space, and the geographies of power*. New York City: Guilford Press. doi: 10.1177/030913259602000113

Boxall, P., & Purcell, J. (2011). *Strategy and human resource management*, 3rd ed. London: Palgrave McMillan Ltd.

Bréger, T. (2019). *Pour une action des collectivités territoriales en faveur de la transition écologique des systèmes alimentaires* [Towards an action of local authorities in favour of the ecological transition of food systems]. Droit rural n°469, étude 3.

Cantillon, G. (2022). Marchés publics et développement durable [Public procurement and sustainable development]. *JCL Environnement et développement durable*, Fasc. 2340.

Champeil-Desplats, V. (2014). *Méthodologie du droit et des sciences du droit*, 1ère éd. Paris: Dalloz, 340.

Chiffoleau, Y., & Prévost, B. (2012). Les circuits courts, des innovations sociales pour une alimentation durable dans les territoires [Short food supply chains as social innovations for sustainable food in territories]. *Norois*, 224, 7–20. doi: 10.4000/norois.4245

Clamour, G. (2014). *La commande publique et l'économie sociale et solidaire* [Public procurement and the social and solidarity economy]. CMP n°10, comm. 248.

Clamour, G. (2018). *Alimentation saine et marchés publics de restauration collective* [Healthy food and public catering contracts]. CMP n°12, comm. 258.

Delzangles, H. (2023). *Vers une commande publique et des acheteurs écologiquement responsables?* [Towards environmentally responsible public procurement and procurers?]. CMP n°3, étude 1.

Dragos, D., & Neamtu, B. (2013). Sustainable public procurement: life-cycle costing in the new EU Directive proposal. *European Procurement & Public Private Partnership Law Review*, 8(1), 19–30. doi: 10.21552/EPPPL/2013/1/159

Dyens, S. (2016). Définition préalable du besoin et sourcing, la responsabilisation des acheteurs publics dans le nouveau droit des marchés publics [Prior definition of the need and sourcing, the empowerment of public buyers in the new public procurement law]. *AJCT*, 422.

European Commission. (2019, 11 December). *The European green deal, COM(2019) 640 final*. Retrieved from https://eur-lex.europa.eu/legal-content/EN/TXT/?uri=COM%3A2019%3A640%3AFIN

European Commission. (2020, 20 May). *A farm to fork strategy, COM(2020) 381 final*. Retrieved from https://eur-lex.europa.eu/legal-content/EN/TXT/?uri=CELEX%3A52020DC0381

European Commission. (2022, 9 June). *Draft council conclusions on the development of sustainable public procurement (2022/C 236/02)*. Retrieved from https://eur-lex.europa.eu/legal-content/EN/TXT/?uri=CELEX:52022XG0620(01)R(01)

Forest, P. (2009). *Approvisionnements transfrontaliers locaux en eau potable entre le Canada et les Etats-Unis: reconsidérations sur le thème de transfert d'eau* [Local transboundary drinking water supplies between Canada and the United States: reconsiderations of the water transfer theme]. Québec: Université Laval. Retrieved from https://corpus.ulaval.ca/entities/publication/1002cfc8-baf0-446a-95e4-93183d9834be

French Mayors' Association. (2016). *Vadémécum encourager l'approvisionnement local* [Encouraging local supply Vademecum]. Retrieved from www.amf.asso.fr/documents-vademecum-encourager-lapprovisionnement-local/14640

Godfray, H. C. J., Aveyard, P., Garnett, T., Hall, J. W., Key, T. J., Lorimer, J., . . . & Jebb, S. A. (2018). Meat consumption, health, and the environment. *Science*, 361(6399), eaam5324. Retrieved from www.science.org/doi/10.1126/science.aam5324

Gosselin, C. F. J. (1908). *Les cantines scolaires* [School canteens]. Paris: Ollier Henri.

Grossfeld, B. (1984). Geography and law. *Michigan Law Review*, 82, 1510–1519. Retrieved from https://repository.law.umich.edu/cgi/viewcontent.cgi?article=3585&context=mlr

Hasquenoph, I. (2021). Sustainable public procurement and geography. *PPLR*, 30(2), 63–77.

Hoepffner, H. (2016). La modification des contrats [Modification of contracts]. *Revue francaise de droit administratif*, 2, 280.

Hoepffner, H. (2018). Le pouvoir de modification et de contrôle [The power of amendment and control]. L'exécution des contrats administratifs [The execution of administrative contracts]. *Le Moniteur*, 65.

Kelly, S., & Swensson, L. F. J. (2017). *Leveraging institutionnal food procurement for linking small farmers to markets*. Rome: FAO.

Kiss, K., Ruszkai, C., & Takács-György, K. (2019). Examination of short supply chains based on circular economy and sustainability aspects. *Resources*, 8, 161. doi: 10.3390/resources8040161

Llorens, F., & Zimmer, W. (2014). *L'instrumentalisation du droit de la commande publique* [The instrumentalization of public procurement law]. CMP n°6, repère 6.

Maréchal, G. (2008). *Les circuits courts alimentaires: bien manger dans les territoires* [Short food circuits: eating well in the territories]. Dijon: Educagri Editions.

National Observatory. (2021). *Observatoire national de la restauration collective bio et durable Résultats de l'enquête 2021*. [National observatory of organic and sustainable collective catering, survey results 2021]. Retrieved from https://observatoire-restauration-biodurable.fr/

Nguyen, H. (2018). *Sustainable food systems (No. CA2079EN)*. Retrieved from www.fao.org/3/ca2079en/CA2079EN.pdf

Nourrisson, D. (2004). Manger à l'école: une histoire morale [Eating at school: a moral story], *Food & History*, 2(1), 227–240.

Nourrisson, D. (2017). *Une histoire du vin* [A history of wine]. Paris: Perrin.

Nourrisson, D. (2020). Faire et savoir-faire à l'école: l'alimentation à l'écran [Doing and knowing at school: food on screen]. *Les enjeux de l'information et de la communication*, n°20/3A, 117–130.

Parayre, S. (2017). L'école primaire après la libération. La reconstruction de la santé par la nutrition, les activités physiques et la médecine scolaire [Primary schools after the liberation. Rebuilding health through nutrition, physical activities and school medicine]. *Fondation Seligmann*, (42), 22–23.

Paris. (2021, 2 September). *Une rentrée toujours plus bio et durable dans la restauration scolaire* [An ever more organic and sustainable return to school catering]. Retrieved from www.paris.fr/pages/pour-la-rentree-quoi-de-bio-dans-la-restauration-scolaire-18393

Pech, A. (2021). *Quand notre environnement nous rend obèses: comment l'environnement influence-t-il nos pratiques alimentaires?, Géoconfluences*. Retrieved from http://geoconfluences.ens-lyon.fr/informations-scientifiques/dossiers-thematiques/geographie-de-la-sante-espaces-et-societes/articles-scientifiques/obesite

Perrier, M. (2017, 9 October). *Quand la municipalité cuisine bio et local* [When municipal cuisine is organic and local]. Retrieved from www.alternatives-economiques.fr/municipalite-cuisine-bio-local/00080820

Reseco. (2019). *Amelioration de la qualite de l'eau en utilisant le levier des achats publics responsables* [Amelioration of the quality of the water using the levier of responsible public purchasing]. Retrieved from https://reseco.fr/wp-content/uploads/2019/09/Qualite_eau_denrees_alimentaires_Rennes_RennesMetropole_2015.pdf

Richer, L. (2015). *La concurrence concurrencée* [Competition challenged]. CMP n°2, étude 2.

Richer, L., & Cantillon, G. (2017). *Achat public et économie des territoires* [Public procurement and the local economy]. Issy-les-Moulineaux: LGDJ.

Robin, Y. (2017). *Food supply procurement, the influence of politics on food supply chains and the governance of local public food service*, Economics and Finances, Université Paris I, 82.

Roux, C. (2019). *Les nouvelles tables du service public* [The new public service tables]. DA n°2, alerte 17.

Stein, A. J., & Santini, F. (2021). The sustainability of "local " food: a review of policy makers, *Review of Agricultural, Food and Environmental Studies*, 103, 77–89. doi: 10.1007/s41130-021-00148-w

Swensson, L. F. J., & Tartanac, F. (2020). Public food procurement for sustainable diet and food systems: the role of the regulatory framework. *Global Food Security*, 25, 100366. doi: 10.1016/j.gfs.2020.100366

Vaysse, C. (2021). *Loi Egalim, AGEC, Climat et résilience, chaud devant pour la commande publique!* [Egalim, AGEC, Climate and resilience Laws, hot on the public procurement front!]. JCPA n°51–52, 2386.

Villeneuve, P. (2018). Droit à la cantine et approvisionnement local en restauration collective: vers de nouvelles obligations pour les collectivités territoriales [Right to canteen and local procurement in collective catering: towards new obligations for local authorities]. *AJCT*, 17.

6 Sustainable public food procurement in Spain

José Pernas García and Ximena Lazo Vitoria

Introduction: The legal framework for sustainable public food procurement in Spain

In this chapter, we aim to analyse the legal framework for sustainable public procurement of food in Spain.

We will outline the provisions of the Public Sector Contract Law.

We will consider a difficulty created by the administrative courts – the legal controversy over the relationship of sustainability clauses to the subject matter of the contract.

We will highlight the lack of technical guidance at national level and contrast that with detailed guidance available particularly from two regions – Valencia and the Basque Country.

We will examine the extent to which legislation on climate change, food waste and rural depopulation has encouraged the purchase of food products with sustainability characteristics.

We will show how procurement of local products has been encouraged by environmental laws passed by six Spanish regions. In most cases the legislation does not include any clear definition of local products.

We will describe the results of a survey of 170 school canteen tenders from five Spanish regions – showing the extent to which they encourage procurement of local food.

Finally, we will consider ways in which sustainable public food procurement could be encouraged – principally through the provision of detailed technical guidance at national level plus a legal requirement within the national procurement legislation requiring minimum environmental standards to be included within all public food procurement.

General assessment of the strategic use of public procurement in law 9/2017

Sustainable public procurement of food in Spain is provided for within the general public procurement framework.[1] Law 9/2017, of 8 November, on Public Sector Contracts of Spain states that

DOI: 10.4324/9781003393023-8

> in all public procurement, social and environmental criteria shall be incorporated in a cross-cutting and mandatory manner whenever they are related to the object of the contract, in the conviction that their inclusion provides better value for money in the contractual provision, as well as a more efficient use of public funds. Likewise, access to public contracting will be facilitated for small and medium-sized companies, as well as for social economy companies.
>
> (Public Sector Contracts Law, 2017, Art. 1.3)

Contracting authorities have wide discretion to integrate these criteria within the procurement exercise – in the way they draw up the detailed tender provisions – technical specifications, selection criteria, award criteria and special performance conditions. This has to take into account the particularities of each service, the capacity and knowledge of the public sector and the market situation.

Lack of technical guidance at national level

However no systematic development of technical criteria has taken place at the national level. The only attempt to standardise criteria at the national level has been carried out through the Green Public Procurement Plan (2018). It contains general objectives for the procurement of services or supplies in the area of "food and catering services". These are of little technical and legal use in the processes of preparing specific contracts.

The advancement of the sustainable food procurement agenda is not a universal development. It may or may not have political support among politicians responsible for public procurement. It depends greatly on the motivation and technical knowledge of the staff of each contracting authority.

There is a legal-administrative culture which is resistant to introducing sustainable food procurement policies.

Despite the legal provisions permitting procurers to give weight to environmental and social criteria, there is still a widespread tendency to ignore such matters and seek only the cheapest suppliers.

Spanish legislation needs to change in harmony with EU law and the new political-legal trend in the EU towards a mandatory approach to green public procurement.

The general legal, political and cultural framework limits the progress of sustainable public food procurement. A legal and technical development of mandatory green public procurement is needed to achieve strategic environmental objectives. The model of the *do no significant harm* (DNSH) principle should be a basis for policy (Pernas García, 2022).

The only reference in the Public Sector Contract Law to public food procurement is found in Article 127. It establishes that when contracting bodies intend to acquire supplies or services with environmental or social characteristics, they may require a label as a means of proof of these characteristics, such as those related to organic agriculture or Fairtrade.

At the regional level, the recent approval of Law 11/2023 of 30 March on the strategic use of public procurement in the Autonomous Community of Aragon stands out. This regulation incorporates the requirement to prioritise "the purchase of seasonal products, fresh food, as well as products of organic farming". Contracting bodies must incorporate

> award criteria that assess the environmental footprint of the contract, taking as a reference the circumstances relating to the production and transport of food, such as the use of short distribution circuits or the existence of compensatory measures that neutralise the negative ecological impact linked to production and transport through environmental investments.
>
> (Aragon, 2023)

The Aragonese legislation established that award criteria could include compliance with Fairtrade or UTZ certification[2] or equivalent sustainable production certifications. The Aragonese regulation also envisaged that catering service contractors should be required to adopt measures against food waste during the execution of their contracts (Art. 33).

Legal controversy: Subject matter of the contract

One of the elements that has dissuaded public sector entities from applying green purchasing in Spain is the legal controversy over the link with the subject matter of the contract. In other words, whether a stipulation made within the tender is relevant enough to the nature of the product or service which is being purchased to justify its inclusion.

EU Directive 2014/24 provided for a broad interpretation of what is meant by the link to the subject matter of the contract (Art. 67.3). Twenty years ago, in the Wienstrom case the European Court of Justice accepted the possibility of using environmental clauses or criteria linked to "processes or production methods", with no link to the operational or functional performance of the contract (CJEU, 2003). In the Concordia case the Court permitted tenders to include clauses that did not generate direct economic return for contracting authorities (CJEU, 2002).

However, despite the rulings made by the European Court of Justice, in Spain the Central Administrative Court of Contract Appeals (TARC) has issued judgments unfavourable to the use of public procurement to support social or environmental objectives. The most important ruling is Resolution 235/2019 (Central Contract Appeals, 2019). The Court found that it was not legally permissible to incorporate strategic criteria within a tender that did not affect the operational quality of the service but that only referred to its social or environmental performance. A strict application of this doctrine would not allow the use of a bid evaluation criterion that referred, for example, to the greenhouse gas emissions of buses in a municipal transport service, the renewable origin of the energy supplied or the sustainable origin of the wood to be used in a public works project. The criteria given as an example have no impact on the quality of the service (wood of conventional origin and wood of sustainable origin offer the same operational performance), they only make it possible to assess the environmental performance of the service.

Similarly, this doctrine would make it illegal for a tender to specify fair trade coffee in a tender since Fairtrade coffee may be indistinguishable in its physical and taste qualities from other coffee.

This doctrinal stance has generated a climate of legal uncertainty, which makes contracting authorities reluctant to include social or environmental provisions within tenders. This uncertainty will continue until the Spanish Parliament passes legislation to amend the Spanish Sector Contracts Law to clarify that European Court of Justice rulings in favour of social and environmental criteria within public tenders have been incorporated into Spanish law.

At least one administrative court of contractual appeals – that of Catalonia – has declared the validity of an award criterion referring to vehicle emission levels, with no impact on the functional performance of the service. The court ruled that certain environmental criteria do not necessarily have to directly improve the service but may promote more general environmental improvement (Catalan Contracts Tribunal, 2020) For a similar ruling see Canaries Contract Tribunal (2022).

Policy developments: Climate change and Waste/circular economy

Spanish environmental policy has been developed in two main thematic areas: climate change and waste/circular economy. Environmental regulations have been developed in respect to these themes at both national and regional level. These regulations, as shown later, include provisions on the use of public food procurement to support environmental policy objectives.

Climate change act

Law 7/2021 of 20 May on climate change and energy transition (hereinafter referred to as the Climate Change Act, 2021) stipulates that emission reduction and carbon footprint criteria specifically aimed at combating climate change will be incorporated as specific technical requirements within public procurement. The Act provided that, within one year, the Ministries of Ecological Transition & Demographic Challenge and Finance should draw up a catalogue of services whose procurement would take into account emissions reduction and carbon footprint, including those related to sustainable and healthy food (Climate Change Act, 2021, Art. 31.1). At the time of writing (September 2023) these criteria have not yet been developed. As and when legally binding sustainable food criteria are developed, based on the best scientific knowledge and adapted to the economic, social and business context, this will do a great deal to promote sustainable public food procurement.

The Climate Change Act (2021) also provided that the specific administrative clauses and terms of food procurement contracts could encourage purchase of food from local suppliers through

> special performance conditions that give priority to fresh or seasonal food, and with a short distribution cycle [*un ciclo corto de distribución]*, provided

> that this is in accordance with the provisions of article 202 of Law 9/2017 of 8 November on Public Sector Contracts.
>
> (Art. 22)

The Climate Change Act does not therefore introduce an obligation, but is merely an expression of the legislature's political intention, which may or may not be implemented by the contracting authorities. There is no further clarification of the concept of the short distribution cycle, which would make this more useful in the contract preparation process.

Regional legislation on climate change also contains generic references to public procurement including criteria or clauses indicating the organic and local origin of food products, promoting seasonal produce or reducing food waste. See, for example, Andalusia (2018); Balearics (2019b); Canary Islands (2022); Navarre (2022) and Valencia (2022b).

Waste/circular economy legislation

Law 7/2022 of 8 April on waste and contaminated land for a circular economy (Waste Act, 2022) envisages a 50% reduction of food waste at both retail and consumer levels. The Waste Act requires that local authorities should adopt food waste prevention programmes with preventive measures aimed at avoiding food waste and encouraging responsible consumption. They should develop ways of utilising surplus food through social initiatives such as soup kitchens or food banks. Prevention of food waste in public kitchens will improve the financial sustainability of local services.

Regional legislation promoting the circular economy has in some cases put particular emphasis on sustainable public food procurement.

An excellent example is Law 3/2023, of 30 March, on the Circular Economy of Andalusia, which provides that

> the reduction of food waste production and the purchase of fresh, seasonal products or products with a short distribution cycle will be promoted in contracts relating to hotel, catering and restaurant services, as well as in food supply contracts, especially in educational, socio-health and health centres, with the establishment of agreements with social entities for the donation of surpluses to be considered.
>
> (Andalucia, 2023, Art. 12)

The corresponding regional legislation in Valencia provides for the development of sustainable public procurement criteria for "reduction of food waste, promotion of food procurement under Mediterranean diet criteria, proximity and seasonal production, sustainable environmental management of food production and contribution to agricultural and marine biodiversity" (Valencia, 2022a, Art. 34.1). See similar measures in Balearics, 2019a and Galicia, 2021b.

Rural depopulation

Initiatives aiming to combat rural depopulation are an important dimension of Spanish social policy. There are a number of regional laws against depopulation which contains provisions encouraging sustainable public procurement of foodstuffs. For example, Law 2/2021, of 7 May, on Economic, Social and Tax Measures against Depopulation and for the Development of the Rural Environment in Castilla-La Mancha provides for measures to use public procurement as a way to promote "local purchases, the use of local, organic, fresh and seasonal products, as well as regionally distinct, high-quality products, which in turn, help to reduce greenhouse gas emissions" (Castilla La Mancha, 2021). Similar provisions can be found in Galicia and Extremadura. Galicia (2021a, Art. 17) provides for the promotion of tender clauses or criteria referring to "fair trade, organic agriculture and livestock". Extremadura (2022, art 56) provides for the promotion of the procurement of "seasonal products", as well as the development of actions to promote "healthy eating based on seasonal and local products".

Public procurement and local products[3]

Debate around local products

Debate surrounding the preference for local products has grown in Spain as elsewhere (Fernández Acevedo, 2022). Buying locally can be seen as a way to reduce GHG emissions through transport. The climate crisis has highlighted the need to reduce the environmental footprint of the food system given its substantial contribution to overall GHG emissions (Crippa et al., 2021).

The experience of food shortages resulting from the COVID-19 pandemic has also aroused greater interest in local food procurement. More recently the Ukraine war and resulting food shortages and price rises has reinforced this tendency.

Regional environmental legislation encouraging local food procurement

Spain's Public Sector Contract Law adopts a green approach, seen for example in its stress on the importance of GHGs in the regulation of award criteria and special performance conditions. However, this 2017 law makes no mention of support for local products. Rather it is climate legislation approved over the last five years, particularly at the level of the autonomous regions, that has paid attention to this matter and, subsequently support for local products has been gradually incorporated into various rules in the Spanish legal system.

Andalucia

In Andalucia Law 8/2018 stipulates that for catering and restaurant services contracts, as well as food supply contracts (especially in educational and health centres) that: "award criteria which emphasize the organic and local origin [*el origen ecológico y de proximidad*] of products and production processes shall be promoted" (Andalusia, 2018, Art. 30.e).

Aragon

The expression "short supply circuit" [*circuito corto de distribución*], has been incorporated into the recent legislation of the Autonomous Region of Aragon. Article 33 of Law 11/2023, of 30 March, on the strategic use of public procurement in the Autonomous Region of Aragon, includes promotion of sustainable food. Its first paragraph states that:

> the procurement authorities shall incorporate award criteria that assess the environmental footprint of the contract, taking as a reference the circumstances relating to the production and transport of the food, such as the use of short supply circuits or the existence of compensatory measures that neutralise the negative ecological impact arising from production and transportation through environmental investments.
>
> (Aragon, 2023)

Asturias and Basque Country

The Law of the Principality of Asturias 1/2023, of March 15, on Environmental Quality, states that the Principality's procurement authorities and other public sector entities in Asturias may include environmental criteria within public tenders, in accordance with the requirements of the legislation on public sector contracts. Local and organic products may be promoted within these clauses, provided that there is a link with the subject matter of the contract and compatibility with community law (Asturias, 2023, Art. 20.2).

Identical wording is found in equivalent legislation which has been adopted in the Basque Country (Basque, 2021, Art. 84.2).

Canaries

Canary Islands' Law 6/2022, of December 27, on climate change and the energy transition encourages public bodies to purchase:

> Products of proximity: principle of food production proximity, extending to other products and services. This makes reference to an indicator of distance between the place of production and the place where it is consumed, which coincides with the territorial scope of the Canary Islands, to reduce pollution from transport and boost the local economy.
>
> (Canary Islands, 2022)

The Canary Islands' legislation links products of proximity primarily to food. However, the principle of *food* production proximity can be extended to other (non-food) products and services. Therefore, a product of proximity shall be any product that complies with the "indicator of distance" established by the regulation, that is, within the Canary Islands.

Navarre

Navarre Law 14/2018, of 18 June, on waste and its taxation includes provisions regulating procurement. Article 19 stipulates that public purchasers should set progressive targets for foods of organic origin or locally-sourced. It requires the inclusion of minimum percentages of foods of organic origin or of proximity as a performance obligation (Navarre, 2018).

Navarre Regional Law 4/2022, of March 22, on climate change and energy transition (Navarre, 2022) advocates "promoting specifically agricultural and livestock products and products of proximity" and "promoting healthy, wholesome food, with seasonal products of high-quality and of proximity".

Terminology: Local products

The overview provided in the previous section shows that legislators in different Spanish regions have made considerable efforts to promote public procurement of local food.

Three different expressions have been used which would refer to the same reality: products of proximity [*productos de proximidad*], local products [*productos locales*] and products from short circuits [*product de circuito corto*]. However, the meaning of these expressions has yet to be defined, even in broad terms, whereas clearer definition would undoubtedly encourage their incorporation into procurement documents.

In the case of "product of proximity", the only legal definition within Spanish procurement law is found within the previously-quoted Canary Islands' Law 6/2022, on climate change and the energy transition which defined "product of proximity" as those produced within the territorial scope of the Canary Islands, which would reduce transport pollution and support the local economy.

Products of proximity in school canteen tenders

Our research examined a representative sample of 170 tenders issued during the years 2018 to 2022. The justification for selecting this period was that 2018 was the year when the new Spanish law on contracts entered into force, based on the European Procurement Directives of 2014. The sample included all public tenders relating to catering for school canteens in five autonomous regions: Castile-La Mancha, the Principality of Asturias, Madrid, Extremadura and the Basque Country.

The research found that 43 of the total 170 tenders studied (25.3%) had clauses relating to local products. Particularly this related to catering contractors being encouraged to purchase local foods.

The preferred option for those clauses was as award criteria – according to which different bids for the contract would be ranked (86%). In a small number of cases (12%) these clauses were included within the technical specification – the detailed description of what the contractor was required to do (12%). The third option chosen was special performance conditions (2%).

There was a wide range of different approaches and significant vagueness in defining what was meant by local product or product of proximity in the contract documents analysed.

Thus, there are specifications that refer to products of proximity, but which do not provide any parameters or concepts. An example of this is the following set of award criteria:

> Technical project (40 points). This shall include the following sections:
>
> Quality of the raw materials used to prepare the menus. Up to 20 points. The following aspects will be evaluated in this section:
>
> The % of organic products and % of fresh products, products of proximity and/or seasonal products used in the preparation of the menus. The points will be distributed as follows:
>
> - higher percentage of organic products in the preparation of the menus: 10 points, allocating the rest proportionally.
> - higher percentage use of fresh products, products of proximity and/or seasonal products in the preparation of the menus: 10 points, allocating the rest proportionally.

This tender specification assigns a score to "products of proximity" but nowhere in the tender documents is this phrase defined or is it made clear how it is to be measured.

In other cases, although a definition of products of proximity is not provided in the tender specifications, it does contain information which makes it possible to ascertain the meaning of this term:

> Origin of seasonal vegetables and to come from crop areas that allow for their transport without first being subjected to freezing processes and within a period of 48 hours prior to their use. Up to a maximum of three points, with the score assigned pro rata according to the time elapsed from product procurement to its use:
>
> From 0 to 24 hours: 3 points
> From 24 to 48 hours: 1.5 points

Finally, there are cases where the contracting authority clearly defines what they would consider to be a product of proximity, defining geographical limits for the origin of production and justifying the need to include this as an assessment criterion:

> Procurement of products of proximity. Based on the closeness of the suppliers of fruits, vegetables, bread and dairy products. This criterion is intended to promote procurement by the contractor of the products indicated provided that the production centre of the products is within a radius of 100 kilometres of the council of Llanera. The aim is to encourage procurement of these categories of products within the aforementioned radius, as this entails less transport, handling and packaging. Energy expenditure in long-distance transport involves greater gas emissions. This means that procurement of these products within a limited km range would reduce energy consumption and the use of complex distribution and storage systems, thereby reducing consumption of packaging and containers for their transport, demand for oil and energy consumption, as well as environmental impact.
>
> This criterion will be scored as follows:
>
> - Fresh seasonal fruits: 6 points
> - Fresh seasonal vegetables: 6 points
> - Bread: 6 points
> - Dairy products: 6 points

Guidance on sustainable food procurement – Valencian Community and Basque Country

Certain regional governments have produced detailed guidance advising procurement officers how to carry out sustainable food procurement. Two notable examples are the Basque region and the Valencian Community.

The Basque region has produced a 43-page practical guide on the use of environmental certifications in public procurement and contracting. This examined in considerable depth the use of eco-labels in public procurement, which eco-labels were most useful and why, how does the Spanish public procurement law provide for the use of environmental management systems, which are the most useful systems and what are common errors in the introduction of environmental criteria in public procurement (Basque, 2020).

The Valencian Community – the autonomous region around Valencia (Valencia, 2018) has produced an 80-page guide which gives detailed guidance for procurement officers on how to carry out public procurement of food and catering services with a focus on specific quality and sustainability criteria.

The guide showed procurement officers how to draw up tender specifications which set out minimum mandatory criteria which would be binding on the company which won the tender. The guide provided a great deal of practical advice. Procurement officers were cautioned that they needed to plan public procurement well in advance to avoid one of the endemic evils of public procurement – lack of time and its derivative: rush. Valencia had introduced a minimum environment requirement that all contracts should have a minimum of 3% organic food, Procurement officers were advised to assess whether suppliers of local or organic food were present in sufficient number, or with the necessary productive capacity to supply the contract, or with adequate logistics and whether they were interested in participating in the tender.

Catering contractors were required to design menus that varied depending on the time of year to incorporate fresh and seasonal foods. Tenders could include a list of seasonal products according to the geographical location in which the contract would be provided. Caterers were required to provide a minimum 40% of fresh, seasonal fruit and vegetables and would receive additional points in the tender assessment the higher the percentage provided.

Draft national legislation on Prevention of Food Loss and Waste

At national level, the draft bill on Prevention of Food Loss and Waste was published in June 2022. This says that public administrations together with the hospitality sector and other food service providers may promote the incorporation of sustainable purchasing criteria (such as fresh, seasonal, local or organic foods) to reduce their environmental footprint and promote the local economy (Food Waste Bill, 2022, Art. 12). The Bill had not become law at the time of writing – September 2023.

Conclusions and recommendations

The Spanish legal regime for public procurement provides opportunities to incorporate environmental and social criteria in public procurement, including food procurement.

Spanish legislation on climate change, waste and circular economy and combating rural population policies, envisages sustainable public procurement as an important policy tool.

The promotion of green or sustainable public procurement in Spain is inhibited by deficiency at a technical level – in that government has not provided public procurers with explicit and detailed technical guidelines in how to implement sustainable public procurement – which need to be sector specific. Separate guidelines are needed at national level for the food and catering sector These could to a large extent be modelled on the aforementioned guidance which has been developed by the Basque and Valencia Regions.

The Public Sector Contract Law should require the drawing up and publication of detailed and sector-specific technical guidelines in how to implement sustainable public procurement, which will develop and adapt the common EU criteria to our reality. The law should further require that these technical guidelines must be taken into account in the contract preparation process and should be the basis for environmental and social clauses for specific tenders. Procurement officers would retain a degree of discretion to take into account the specific characteristics of the service or supply to be contracted, local market conditions, local capacity of small businesses. Certain minimum technical criteria could be applied in the contractual preparation process – such as a minimum percentage of organic food.

There exists a clear trend in Spanish legislation in favour of incorporating clauses on local products into public tenders. This occurred initially in climate legislation and has subsequently extended to other regulatory sectors.

This trend occurs within the framework of the European Green Deal and its "Farm to Fork" Food Strategy and is motivated by the urgent need to reduce GHGs and contribute to attaining the decarbonisation targets set under the Paris Agreement.

Our analysis of Spanish regulations revealed use of the following three legal concepts: "product of proximity", "local product" and "short supply chain". However, we have criticised the lack of legal definitions for these terms, with some exceptions such as the case of Canary Islands' legislation.

The practical data presented earlier testify to the gradual incorporation of this element into specifications for the school canteens studied and this is an indication of what may be happening in other similar services. A very wide range of different clauses were found, and it was possible to understand the concept of proximity in a clear way in a limited number of cases.

Notes

1 This section was written by Jose Pernas García. This study is part of the Research Project "Local circular economy: public policies, legal tools and best practices in circular municipal waste management", Reference: TED2021–129734B-I00, financed by the Spanish Ministry of Science and Innovation.

2 UTZ Certified is a programme and a label for sustainable farming. www.rainforest-alliance.org/utz/.

3 This section was written by Ximena Lazo Vitoria. This study is part of the Research Project "Green Public Procurement: a strategic tool for tackling climate change. An analysis of the food, building and sustainable transport sectors, as well as in the context of rural depopulation", Reference: PID2021–122131OB-I00, financed by the Spanish Ministry of Science and Innovation.

Bibliography

Academic articles – English

Crippa, M. Solazzo, E., Guizzardi, D., Monforti-Ferrario, F. T., Ubiello, F., N., & Leip, A. (2021). Food systems are responsible for a third of global anthropogenic GHG emissions. *Nature Food*, 2, 1–12. doi: 10.1038/s43016-021-00225-9

Academic articles – Spanish

Fernández Acevedo, R. (2022). *Los criterios de adjudicación al servicio de la calidad ambiental y de la lucha contra el cambio climático: el precio en jaque* [Award criteria in the service of environmental quality and the fight against climate change: price in check], in Lazo Vitoria, X. (ed.), *Compra pública verde y cambio climático*. Barcelona: Atelier.

Pernas García, J. J. (2022). *El principio DNSH del mecanismo de recuperación y resiliencia y su incidencia en la gestión contractual del plan de recuperación, transformación y resiliencia* [The DNSH principle of the recovery and resilience mechanism and its impact on the contractual management of the recovery, transformation and resilience plan], in De Guerrero Manso, C., Gimeno Feliu, J. M. (eds.), *Observatorio de los contratos públicos 2021*. Cizur Menor (Navarre): Aranzadi.

List of national laws

Climate Change Act. (2021). *Ley 7/2021, de 20 de mayo, de cambio climático y transición energética* [Law 7/2021 of 20 May on climate change and energy transition]. Retrieved from www.boe.es/buscar/act.php?id=BOE-A-2021-8447

Food Waste Bill. (2022, 17 June 2022). *Proyecto de Ley de prevención de las pérdidas y el desperdicio alimentario* [Food loss and waste prevention bill]. Official Journal of the Spanish Parliament. Retrieved from www.congreso.es/public_oficiales/L14/CONG/BOCG/A/BOCG-14-A-107-1.PDF

Green Public Procurement Plan. (2018). *Orden PCI/86/2019, de 31 de enero, por la que se publica el Acuerdo del Consejo de Ministros de 7 de diciembre de 2018, por el que se aprueba el Plan de Contratación Pública Ecológica de la Administración General del Estado, sus organismos autónomos y las entidades gestoras de la Seguridad Social* [Green public procurement plan for the general state administration, its autonomous bodies and social security management bodies (2018–2025), approved by order PCI/86/2019 of 31 January]. Retrieved from www.boe.es/diario_boe/txt.php?id=BOE-A-2019-1394

Public Sector Contracts Law. (2017). *Ley 9/2017, de 8 de noviembre, de Contratos del Sector Público* [Law 9/2017, of 8 November, on public sector contracts of Spain]. Retrieved from www.boe.es/buscar/act.php?id=BOE-A-2017-12902

Waste Act. (2022). *Ley 7/2022, de 8 de abril, de residuos y suelos contaminados para una economía circular* [Law 7/2022 of 8 April on waste and contaminated land for a circular economy]. Retrieved from www.boe.es/buscar/act.php?id=BOE-A-2022-5809

List of regional regulations and other documents

Andalucia. (2023). *Ley 3/2023, de 30 de marzo, de Economía Circular de Andalucía* [Law 3/2023, of 30 March, on the Circular Economy of Andalusia]. Retrieved from www.boe.es/buscar/act.php?id=BOE-A-2023-9957

Andalusia. (2018). *Ley 8/2018, de 8 de octubre, de medidas frente al cambio climático y para la transición hacia un nuevo modelo energético en Andalucía* [Law 8/2018, of 8 October, on measures against climate change and for the transition to a new energy model in Andalusia]. Retrieved from www.boe.es/buscar/doc.php?id=BOE-A-2018-15238

Aragon. (2023). *Ley 11/2023, de 30 de marzo, de uso estratégico de la contratación pública de la Comunidad Autónoma de Aragón* [Law 11/2023 of 30 March on the strategic use of public procurement in the Autonomous Community of Aragon]. Retrieved from www.boe.es/buscar/act.php?id=BOE-A-2023-10876

Asturias. (2023). *Ley 1/2023, de 15 de marzo, de Calidad Ambiental* [Law 1/2023, of March 15, on Environmental Quality]. Retrieved from www.boe.es/diario_boe/txt.php?id=BOE-A-2023-10347

Balearics. (2019a). *Ley 8/2019, de 19 de febrero, de residuos y suelos contaminados de las Illes Balears* [Law 8/2019 of 19 February 2019 on waste and contaminated soils of the Balearic Islands]. Retrieved from www.boe.es/buscar/doc.php?id=BOE-A-2019-5577

Balearics. (2019b). *Ley 10/2019, de 22 de febrero, de cambio climático y transición energética de Illes Balears* [Law 10/2019 of 22 February 2019 on climate change and energy transition]. Retrieved from www.boe.es/buscar/doc.php?id=BOE-A-2019-5579

Basque. (2020). *Guía práctica sobre el uso de las certificaciones ambientales en la compra y contratación pública* [Practical guide on the use of environmental certifications in public procurement and contracting]. Retrieved from www.euskadi.eus/contenidos/documentacion/guia_certificaciones/es_def/adjuntos/Guia-certificaciones-ambientales-ihobe-cast.pdf

Basque. (2021). *Ley 10/2021, de 9 de diciembre, de Administración Ambiental de Euskadi* [Law 10/2021, of December 9, on Environmental Administration in the Basque Country]. Retrieved from www.boe.es/buscar/act.php?id=BOE-A-2022-951

Canary Islands. (2022). *Ley 6/2022, de 27 de diciembre, de cambio climático y transición energética de Canarias* [Law 6/2022, of 27 December, on climate change and energy transition in the Canary Islands]. Retrieved from www.boe.es/buscar/doc.php?id=BOE-A-2023-2941

Castilla-La Mancha. (2021). *Ley 2/2021, de 7 de mayo, de Medidas Económicas, Sociales y Tributarias frente a la Despoblación y para el Desarrollo del Medio Rural en Castilla-La Mancha* [Law 2/2021 of 7 May on Economic, Social and Tax Measures against Depopulation and for the Development of the Rural Environment in Castilla-La Mancha]. Retrieved from www.boe.es/buscar/act.php?id=BOE-A-2021-11513

Extremadura. (2022). *Ley 3/2022, de 17 de marzo, de medidas ante el reto demográfico y territorial de Extremadura* [Law 3/2022, of 17 March, on measures to face the demographic and territorial challenge in Extremadura]. Retrieved from www.boe.es/buscar/act.php?id=BOE-A-2022-5853

Galicia. (2021a). *Ley 5/2021, de 2 de febrero, de impulso demográfico de Galicia* [Law 5/2021, of 2 February, on demographic promotion in Galicia]. Retreived from www.boe.es/buscar/act.php?id=BOE-A-2021-5136

Galicia. (2021b). *Ley 6/2021, de 17 de febrero, de residuos y suelos contaminados de Galicia* [Law 6/2021 of 17 February on waste and contaminated soils of Galicia]. Retrieved from https://boe.es/buscar/act.php?id=BOE-A-2021-5137

Navarre. (2018). *Ley Foral 14/2018, de 18 de junio, de residuos y su fiscalidad de Navarra* [Regional Law 14/2018 of 18 June on waste and its taxation in Navarre]. Retrieved from www.boe.es/buscar/doc.php?id=BOE-A-2018-8953

Navarre. (2022). *Ley Foral 4/2022, de 22 de marzo, de Cambio Climático y Transición Energética de Navarra* [Regional Law 4/2022 of 22 March on Climate Change and Energy Transition (Navarra)]. Retrieved from www.boe.es/buscar/act.php?id=BOE-A-2022-6402

Valencia. (2018). *Decreto 84/2018, de 15 de junio, del Consell, de fomento de una alimentación saludable y sostenible en centros de la Generalitat* [Decree 84/2018, of 15 June, of the Consell, on the promotion of healthy and sustainable food in centres of the Valencian Community]. Retrieved from https://dogv.gva.es/portal/ficha_disposicion.jsp?L=1&sig=006177%2F2018

Valencia. (2021). *Comunidad Valenciama, Compra pública de alimentos y de servicios de alimentación con criterios de calidad, sostenibilidad y responsabilidad* [Public procurement of food and food services with criteria of quality, sustainability and responsibility]. Retrieved from https://sostenibilidadalimentaria.org/wp-content/uploads/2021/10/guia-practica-pais-valencia-compra-publica-sostenibilidad-alimentaria-gosa-cerai.pdf

Valencia. (2022a). *Ley 5/2022, de 29 de noviembre, de residuos y suelos contaminados para el fomento de la economía circular en la Comunitat Valenciana* [Law 5/2022 of 29

November on waste and contaminated land for the promotion of the circular economy in the Valencia Region]. Retrieved from www.boe.es/buscar/doc.php?id=BOE-A-2023-3348

Valencia. (2022b). *Ley 6/2022, de 5 de diciembre, del Cambio Climático y la Transición Ecológica de la Comunitat Valenciana* [Law 6/2022 of 5 December on Climate Change and Ecological Transition of the Valencian Community]. Retrived from www.boe.es/buscar/act.php?id=BOE-A-2023-4378

List of cited judgments of the Court of Justice of the EU (CJEU)

CJEU. (2002). *Judgment of 17 September 2002, Case C-513/99, Concordia Bus Finland.* C-513/99, Concordia Bus Finland. Retrieved from https://curia.europa.eu/jcms/jcms/Jo1_6308/

CJEU. (2003). *Judgment of 4 December 2003, Case C 448/01, EVN and Wienstrom.* Retrieved from https://curia.europa.eu/jcms/jcms/Jo1_6308/

List of decisions of administrative tribunals

Canaries Contract Tribunal. (2022). Resolution 31/2022, of 2 February. *Tribunal Administrativo de Contratos Públicos de la Comunidad Autónoma de Canarias* [Administrative Tribunal of Public Contracting of the Canary Islands]. Retrieved from www.gobiernodecanarias.org/hacienda/tacp/index.html

Catalan Contracts Tribunal. (2020). Resolution 349/2020, of 29 October. *Tribunal Català de Contractes del Sector Públic* [Catalan Public Sector Contracts Court]. Retrieved from https://contractacio.gencat.cat/ca/contacte/tccsp/

Central Contract Appeals. (2019). Resolution 235/2019, of 8 March. *Tribunal Administrativo Central de Recursos Contractuales* [Central Administrative Court of Contract Appeals (TARC)]. Retrieved from www.hacienda.gob.es/es-ES/Areas%20Tematicas/Contratacion/TACRC/Paginas/Tribunal%20Administrativo%20Central%20de%20Recursos%20Contractuales.aspx

List of abbreviations

DNSH Do No Significant Harm
ECJ European Court of Justice
CJEU Court of Justice of the European Union
TARC Tribunal Administrativo Central de Recursos Contractuales [Central Administrative Court of Contract Appeals]

7 Sustainable public food procurement in Brazilian schools

An obligation under and beyond the National School Feeding Programme (Programa Nacional de Alimentação Escolar – PNAE)

Chiara Falvo[1]

Introduction

With more than 15 million people employed in more than 5 million agribusinesses (IBGE, 2019), Brazil is one of the world's largest producers and exporters of agricultural products. Food production is very diverse due to the country's vast territory, varied climatic and geographical conditions, and rich biodiversity. Farm structures are also highly diverse: large, export-oriented farms dominate the agricultural sector, but millions of smallholders and family farmers are responsible for most of the food consumed by the Brazilian population (IBGE, 2019). Being one of the countries with the highest land concentration in the world (Luzzi De Campos & Goldfarb, 2021), Brazil is confronted with widespread rural poverty and pesticide-intensive monocultures that hinder the development of more diverse, equitable and sustainable agriculture. It is therefore mainly due to income distribution issues and affordability constraints (OECD/FAO, 2022) that the country still faces significant challenges related to food and nutritional insecurity, also aggravated by the COVID-19 pandemic, rising food inflation and the ongoing Russo-Ukrainian War. Access to healthy and nutritious food remains a major issue, with a high rate of obesity and overweight among adults, especially women and children.

In this context, Brazil's institutional food procurement programmes stand as strategic and comprehensive public policies for combating hunger and poverty and promoting rural and sustainable development. Dating back to 1955, the National School Feeding Programme – *Programa Nacional de Alimentação Escolar* (PNAE) – aims to contribute to children's development and the promotion of healthy eating habits by providing meals and food education in public schools throughout Brazil. The Food Purchase Programme – *Programa de Aquisição de Alimentos* (PAA) –, launched in 2003 in the context of the Zero Hunger strategy, aims to fight food and nutritional insecurity and encourage family farming. This programme marked the beginning of a major governmental intervention to stimulate demand

DOI: 10.4324/9781003393023-9
This chapter has been made available under a CC-BY-NC-ND 4.0 license.

for smallholder production through public procurement, making Brazil the first country to adopt a policy of this kind (Swensson, 2019).

The PAA and the PNAE are complementary programmes, with similar objectives, national reach, and a focus on areas where family farming supply and vulnerable population demand converge (Bravo et al., 2022). Over time, several institutional and regulatory changes have been made to improve the programmes. In particular, preferential procurement schemes (Swensson & Tartanac, 2020) and a wide range of sustainability requirements connected to which food to purchase, from whom to source it and how to purchase it have been included.

Brazil, therefore, represents a very interesting case study for examining how the legal framework may support the strategic use of public food procurement to advance multiple sustainability objectives. The Brazilian experience also shows that, despite the favourable regulatory framework, structural barriers need to be overcome to achieve broad social and environmental goals. Several studies have explored the implementation of specific public food procurement initiatives (Soares et al., 2021; FAO, 2021; Kluczkovski et al., 2022). However the legal mechanisms that enable more equitable, diversified, and sustainable food procurement remain overlooked (Swensson & Tartanac, 2020).

The present chapter aims to fill this gap by analysing selected procurement rules that contribute to social and environmental objectives, supported, to the extent possible, by some findings from interdisciplinary literature. By relying significantly on sources in the original language, this qualitative study also aims at making knowledge about the legal aspects of public food procurement in Brazil available to a wider audience.

The focus of the analysis is on public food purchases under the PNAE, due to the author's research interest in the school environment and the variety of sustainability inputs provided by the programme. Yet, some aspects of the PAA and the overall public procurement framework will be also mentioned given their relevance for the study.[2]

Sustainable public food procurement under the National School Feeding Programme (PNAE)

An outline of the PNAE's evolution, objectives and organisation

With decades of history, Brazil's National School Feeding Programme is one of the largest school feeding programmes in the world, providing around 50 million free meals a day, reaching 40.5 million students and more than 160.000 schools in 5.570 municipalities (Bravo et al., 2022; Luzzi De Campos & Goldfarb, 2021; Schottz & Schmitt, 2021). Created in 1955, the programme was originally centralised, but in the 1990s was restructured to decentralise and divide responsibilities between different government levels (CMAP, 2020)[3] PNAE is funded through a system of financial cooperation among different federal entities. The Brazilian central government pays a budget supplementation to States, municipalities and federal schools based on the number of students enrolled in the public education system, which is to be used exclusively for food procurement.[4]

Law n° 11.947 of 16 June 2009 was a landmark in the evolution of the PNAE, as it established "the first direct and institutional interaction between the PNAE and family farming" (Marques, 2020). This law (2009) (hereinafter L.) contains the programme's general rules.

This law has been extensively regulated by successive instruments. In particular, FNDE (2020) (hereinafter R.) "lays down the rules for the technical, administrative and financial execution of the PNAE, addressed to States, the Federal District, Municipalities and federal schools" (Art. 1 R.), providing for "the consolidation and refinement of standards that were previously scattered in several regulatory instruments" (first recital). FNDE (2020) is the primary source of the analysis that follows.

PNAE's main objective is "to contribute to the bio-psychosocial growth, development, learning and school performance of students and the creation of healthy eating habits, through food and nutritional education and the provision of meals that cover their nutritional needs during the school period" (Art. 4 R).

The main actors involved in the programme are listed in Art. 7 R. They comprise the National Fund for the Development of Education (FNDE) – linked to the Ministry of Education, responsible for coordinating, regulating and funding the programme; the Executing Entities (EEx) – State Education Divisions, municipal governments and federal schools, responsible for planning, managing and implementing the programme, including by using and complementing federal funding, as well as reporting and providing school meals and food education; the School Meals Council (CAE) – a collegiate body established at the State, Federal District and Municipal levels, in charge of monitoring, deliberating and advising on the programme[5]; and, finally, the Executing Unit (UEx) – a private non-profit entity representative of the school community, those responsibilities vary depending on the management model.

Each Executing Entity can define the management model of the programme (centralised, decentralised or mixed) and of the food service (self-managed or outsourced). In the centralised management, the EEx purchases the foodstuffs that are then supplied to school units for preparing and serving the meals. In the decentralised model, the EEx transfers financial resources to the school's UEx, which directly purchases foodstuffs (Art. 8, I, II R.). It is interesting to highlight two points: first, a new system of payment, called PNAE Card Account, will become mandatory for the EEx to receive and use federal funds (Art. 47, XII R.) to improve procurement transparency by identifying suppliers and payments (COEFA, 2020). Secondly, the law sets specific obligations to ensure the viability of the decentralised model, as "the EEx must ensure the necessary structure" for the realisation of the due procurement process, the authorisation of expenses, management and execution of public contracts, the control of stock and storage of foodstuffs, as well as reporting and other measures for the proper use of financial resources (Art. 9 R.).

Finally, the monitoring of the programme is the responsibility of a numbers of governmental and non-governmental actors: the FNDE, the School Meals Councils, the other entities responsible for the education systems, the Office of the Comptroller General (CGU), the Federal Audit Court (TCU), the Public Prosecutor's Office and society in general (Art. 63 R.). In fact, "any individual, association or union, as well as other legal entities that represent society in the control of public administration, has legitimation to denounce irregularities or illegalities in the execution of the PNAE to the FNDE" (Art. 66 R.).

The programme has been organised on the basis of decentralisation, collaboration between different actors and social participation. Decentralisation has improved the rationalisation of logistics and distribution costs (Marques, 2020). It has promoted responsiveness to local realities and a high standard of food provision, while also contributing to the development of local economies through the creation of institutional food demand at the municipal level (CMAP, 2020). Collaboration and social participation are manifest in the programme's decision-making, management and monitoring mechanisms, as well as in the legal framework (Schottz & Schmitt, 2021), thus contributing to the programme's democratisation (Marques, 2020).

In the next sections, sustainability requirements for school food procurement under the PNAE legal framework will be discussed.

Mandatory sustainable public food procurement under the PNAE

In attendance to the fundamental rights to education and food (Arts. 6, 205 and 208, Federal Constitution of 1988, hereinafter CFRB), PNAE affirms the right to (free) school feeding for all students of public basic education (from nursery to high school), and the respective State duty to promote it, observing the guidelines set by the law (Art. 3 R.). These are I – the use of healthy and adequate food; II – the inclusion of food and nutritional education in the school curriculum; III – the universality of the school feeding service; IV – community participation in monitoring of public authorities' agency to ensure the provision of healthy and adequate school meals; V – the promotion of sustainable development, incentivising the acquisition of diversified foodstuffs, produced locally and preferably by family farmers and rural family entrepreneurs, prioritising indigenous and *quilombola* communities (descendants of Afro-Brazilian slaves); and finally, VI – the right to school feeding, ensuring students' equal access to food and nutritional security and respecting biological and socio-economic differences (Art. 5). The acquisition of foodstuffs under the PNAE must observe these guidelines (Art. 13 L.; Art. 23 R.).

"What to buy" requirements

PNAE legislation sets several requirements affecting the definition of the subject matter of the contracts for school food. These requirements prescribe what can be purchased, what must be purchased, and what cannot be purchased.

First, the financial resources transferred by the FNDE under the PNAE can be used exclusively for the acquisition of foodstuffs (Art. 5°, §2° L.; Art. 47, caput; Art. 51, caput R.). Therefore, when outsourcing the catering service, the EEx can use PNAE's funds to pay for meals (Art. 13 R.) or foodstuffs, while other expenses are to be borne with its own resources (Art. 51, §2° R.). In these cases, the EEx must ensure contractors' compliance with the relevant legislation and FNDE guidelines (Art. 13, 1°§ R.), and must ensure that they provide invoices specific to foodstuffs (Art. 27, II R.). Additionally, access to the facilities and the documentation necessary to verify compliance with the contract and the Programme's rules must be guaranteed to the EEx, CAE, FNDE, and other monitoring entities in both the tender notice and the contract (Art. 13, 2°§ R.).

The acquisition of foodstuffs under the PNAE must comply with the menu planned by the responsible nutritionist (RT), following the Resolution's provisions (Art. 23 R.). School menus must be developed by the PNAE nutritionist attached to the EEx, "based on the use of raw or minimally processed food, respecting nutritional needs, eating habits and local food culture, and aiming at sustainability, seasonality, the agricultural diversification of the region and the promotion of adequate and healthy nutrition" (Art. 17 R.). Menus must also be adapted to diagnosed special dietary needs (Art. 17, §1° R.) and to differentiated nutritional demands (Art. 12, §2° L., introduced by Law 12.982/2014), including vegetarian habits (COSAN, 2020b). They must also meet the cultural specificities of indigenous and/or *quilombola* communities (Art. 17 §3° R.).

Students' access to a greater variety of foods is also recommended, as an important indicator of diet quality and a factor that contributes to increased acceptance of the meals provided. The concept of variety is approached in two dimensions: food offered on menus weekly (Art. 19, §1° R.) and food products purchased annually (Art. 21, Parágrafo único, R.) (COSAN, 2020a).

Some requirements concern the minimum nutritional value of meals (Arts. 18–19 R.). Minimum levels of fruit and vegetables, vitamins and iron sources are set. To meet the requirement to provide foods containing heme iron at least four days a week, school managers may not serve vegetarian meals to all students for more than one day per week (COSAN, 2020b).

PNAE funds must be allocated as following: at least 75% must be used to purchase raw or minimally processed foods, a maximum of 20% for processed and ultra-processed foods, and a maximum of 5% for processed culinary ingredients (Art. 21 R.).

Finally, there are stipulations relating to products to be restricted and others prohibited, (Arts. 18, 19). Additionally, PNAE resources cannot be used to purchase a list of ultra-processed foods and beverages to prevent obesity, improve health and build healthier eating habits in the school environment (Art. 22 R; COSAN, 2020a).

The correct specification of the subject matter of contracts, in this case of foodstuffs, is an obligation of the public administration (Art. 15, §7° Law n° 8.666 of 1993; Art. 40, §1° Law n° 14.133 of 2021). This is essential for effective procurement. As reported by the Coordination of Food and Nutritional Security (COSAN), the lack of and/or inaccurate specification of foodstuffs in school feeding purchases constitutes a significant barrier in the purchasing process, leading to issues like delays or failures to satisfy the consumers' needs and higher costs (COSAN, 2016).

For this reason, PNAE nutritionists must participate in the procurement process with regard to the definition of specifications and quantities. Another mandatory technical activity for nutritionists is to engage with family farmers and their organisations to learn about local production (CFN, 2010). The key role of nutritionists in the implementation of the programme is evidenced also by the rule providing for the suspension of the transfer of financial resources to the EEx that has not registered the responsible nutritionist in the FNDE system (Art. 56 R).

It is important to emphasise that the decision to entrust the responsibility for menu design to nutritionists at the EEx level is crucial to allow states and municipalities to maintain their autonomy in defining menus and purchases according to local specificities.

Finally, the Resolution's part "on the food procurement process" opens with two general preferences, which are materialised through a number of rules and mechanisms that will be explored in the following sections. Firstly, the purchase must be made, whenever possible, in the same municipality where the schools are located. Buying local is expected. Secondly, organic and/or agroecological foods must be prioritised (Art. 23 R.).

Who-to-source-from requirements

Since 2009, the PNAE legislation requires part of the food to be sourced from specific groups of suppliers. Reserving procurement opportunities to specific categories of suppliers is a tool to overcome their inability to get access to public contracts under normal conditions of competition (De Schutter et al., 2021). The set-aside mechanism, or reservation scheme (Swensson & Tartanac, 2020), provides that

> at least 30% of the financial resources transferred by the FNDE under the PNAE must be used to purchase foodstuffs directly from Family Farming and the Rural Family Entrepreneur or their organisations, giving priority to land reform settlers, traditional indigenous communities and quilombola communities.
>
> (Art. 14 L. and Art. 29 R.)

If the tied funds are not used for the intended purpose, the EEx must repay them to the FNDE (Art. 29, § 1° R.). The set-aside obligation can only be waived in three circumstances, which must be duly proven (Art. 29, §2° R.).

How to buy food: Different procedures for different purposes

All contracts for the purchase of food under the PNAE are subject to the general public procurement regime (Law 8.666/1993 that will be replaced by Law 14.133/2021 from 30 December 2023). There is also an explicit legal requirement for transparency, as information on the food purchasing process must be broadly publicised (Art. 26 R).

The PNAE legislation prescribes two different procedures for purchasing foodstuffs, depending on the type of supplier. When purchasing from family farmers, the law allows a waiver of bidding (Art. 14, §1°, L.) and, since 2013, subsequent Regulations have prescribed a specific purchase process called public call (Rezende et al., 2022). With other suppliers the procurement must be made through a competitive bidding procedure in the form of an electronic reverse auction (*licitação, modalidade pregão eletrônico*) (Art. 24, II R). This kind of auction is a simpler and faster process compared to the traditional bidding and was created to enable a faster and more efficient procedure for day-to-day purchases by the Public Administration (see Decree n° 10.024 of 2019). Concerning this procedure, PNAE legislation adds two sets of rules to the general discipline, mainly aimed at ensuring proper pricing, transparency and accountability (Arts. 27–28 R.).

The public call: Purchasing from family farmers

Brazilian public food procurement law differs from that of many other countries in the extent to which it supports and encourages purchasing from family farmers.

In fact, under certain conditions, the law allows purchases from family farmers without competitive bidding. Prices must be consistent with those on the local market, food must meet the quality control requirements established by relevant rules (Art. 30 R.) and the constitutional principles of public administration – legality, equal treatment, morality, publicity, transparency and efficiency – must be observed (Art. 37 CFRB).

When these conditions are met, the EEx that opts for waiving the bidding procedure must purchase through a specific process called public call – *chamada pública* or *chamamento público* – which is similar to a request for proposals. This process is defined as "the administrative procedure aimed at the selection of a specific proposal for the purchase of foodstuffs from Family Agriculture and/or Rural Family Entrepreneurs or their organisations" (Art. 30, §1°, §2° R.).

The public call constitutes an exception to the Public Administration's obligation to tender (Art. 37, XXI of CFRB and Art. 2 of Law 8.666/93), which, nonetheless, "reflects the application of the principle of publicity, to the extent that, in a transparent manner, the Administration communicate its objectives and allows interested parties from the private sector to respond to the extent of their interests" (Carvalho Filho, 2019). The procedure was designed to be simpler and faster, taking into consideration that smallholders' participation in procurement procedures could be limited by tendering rules that establish bureaucratic systems (Nehring et al., 2017).

Moreover, support in the implementation of the different phases of the process is provided through model documents (Annexes to FNDE, 2020).

Under the PNAE legislation, a first set of rules that differentiate the two procedures concerns the determination of the reference price of foodstuffs. In the public call, this "should be determined by EEx based on market price research" (Art. 31 R.). It

> must be the average price surveyed in at least three local markets, prioritising family farmers' markets, when available, plus the inputs that are required in

> the public call, such as freight, packaging and any other charge necessary for supplying the products.
>
> (§1°)

"If it is impossible to survey the local level, this must be carried out or complemented at the level of the immediate, intermediate, state or national geographic regions, in this order" (§2°). This is the regional subdivision of Brazil adopted in 2017 by the Brazilian Institute of Geography and Statistics (IBGE), updated for public policy planning and management purposes taking into consideration recent social, demographic, economic and political changes. Several PNAE rules make reference to it (COSAN, 2020c).

Therefore, to establish a purchasing price that considers geographical differences and the specific reality of family farming, the law gives priority to proximity and family farmers' markets and expressly provides an example of the additional charges that need to be added to the market price. A further provision that deserves attention defines that when it is impossible to survey prices for agroecological or organic products, the EEx may increase their prices up to 30% compared to conventional products (Art. 31 § 5). This provision, which is also contained in the PAA (Art. 4°, §1° Law 14.628 of 2023), recognises the existence of distinct markets for environmentally sustainable products and sets the extra cost margin that can be justified due to their positive attributes.

Moreover, "public meetings open to all interested parties, may be held before the launch of public calls, to collect information and clarify any doubts about the process for acquiring family farming products" (Art. 31 § 3). Although these preliminary market consultations are mentioned in relation to the purchase price, their scope seems to encompass any activity of market engagement to align demand and supply and broaden participation.

Finally, "the purchase prices defined by EEx must be indicated in the public call and must be the prices paid to family farmers, rural entrepreneurs and/or their organisations for the sale of the foodstuff" (Art. 31 § 4). Therefore, price would not be a criterion on which suppliers would compete. The abandonment of the lowest price as the main selection criterion for purchases of family farming products is one of the major features of this procedure (Schottz & Schmitt, 2021).

To promote broad participation in the public call, the publication and advertising must be also adapted to the specific audience. The publication must occur in different channels, such as official websites, posters in public places or newspapers and local radio, while the advertising must address local family farming organisations and public technical assistance and rural extension entities. The invitation must remain open for a minimum period of 20 calendar days (Art. 32 R.) and must contain a detailed description of the demanded food items, quantities and logistics of distribution and delivery (Schottz & Schmitt, 2021). Only the items listed in the call may be later accepted as variants, if necessary (Art. 33 R.).

The public call procedure is also subject to special rules regarding the qualification and selection of sales projects. Sale projects may be presented by both formal (associations and cooperatives) and informal groups of producers and individual suppliers, which must provide the relevant documentation for qualification (Arts. 34, 36 R.).[6]

An important advance introduced by Resolution 6/2020 (Art. 35 R.) is the evaluation of sales projects based on two prioritisation categories: territorial and by characteristics of producers (CMAP, 2020). In fact, the project must be ranked from the lowest to the highest territorial level (following the IBGE Regional Division), giving priority to municipal-level production. Within each territorial level, priority must be given to land reform settlements, traditional indigenous communities and *quilombola* communities, then to suppliers of certified organic or agroecological food, and, in the third place, according to the level of organization of farmers, to formal groups, followed by informal groups and individual suppliers.

These rules deserve special attention. As the doctrine points out (Rezende et al., 2022), both the prioritisation criteria and the higher reference price for organic and agroecological products, are in accordance with the school feeding guidelines set by the Law, mainly "the use of healthy and adequate food", "support to sustainable development" and "the right to school feeding".

Moreover, being aligned with the country's geographical panorama, the new model of prioritisation criteria allows for maintaining the relationship between local production and the school environment (COSAN, 2020c). Prioritising family farmers located closer to the municipality where the public call originates allows to strengthen local family farming, purchase fresher food, improve income generation for families and the local/regional economy, reduce the rural exodus, and facilitate the logistics of product delivery (COSAN, 2020c).

This is particularly important for indigenous and *quilombola* communities. In fact, matching the obligation to adapt school menus to the food culture of indigenous and *quilombola* communities (Art. 17, §°3 R) with the preferential purchase of food products directly from these target populations is crucial to overcome structural barriers and guarantee their rights. Conversely, the non-acquisition of traditional foods for the consumption of these populations, often replaced by industrialised food that needs to be transported over long distances, not only prevent their access to culturally contextualised food, but is also associated with logistical and connected health, environmental and budgetary problems, such as irregular provision and low nutritional quality of meals (MPF, 2020). Giving these communities access to the school food market not only improves their food and nutritional security, but also enhances their traditional practices, which form part of Brazil's cultural heritage (Art. 216 CFRB), as well as their role in preserving agrobiodiversity. In this regard, it is important to emphasise that according to the legislation, the food produced by traditional populations, similarly to that intended for household consumption, is exempt from registration, inspection and control. The adaptation of sanitary standards and practices to the production and marketing of food by traditional peoples and communities, including in the context of school feeding, is key to ensure their participation in public procurement procedures (MPF, 2020).

As pointed out by Swensson (2019), through the prioritisation of groups over individuals, the programme encourages producer organisations as a way to support smallholders' compliance with the purchase requirements (such as quality and safety, volumes, regularity, market prices, tax regularisation), as well as their structuring to gain access to other formal markets and to services (such as credit and technical assistance), inputs and information. On the administration side, this can reduce the transaction costs of dealing with many individual producers. The greater supply efficiency of structured farmer organisations (Nehring et al., 2017) could also explain the exception allowing EEx receiving funds above a certain annual threshold to accept proposals only from formal groups (Art. 37 R.).

The PNAE foresees an individual sale limit for each family farmer or rural family entrepreneur and their organisations (Art. 39 R.). This provision is intended to avoid excessive benefits for a family unit or specific organisation from the implementation of this public policy (AGU, 2023). Between 2015 and 2021, the limit was doubled from R$ 20,000 to R$ 40,000,[7] taking into account the impact of accumulated inflation on production costs, which directly affects family farmers, as well as the understanding that providing healthy and adequate school meals involves supporting sustainable development, including the viability of family farming supply and the sustainability of the short chain model, which is a reference within the programme (Rezende et al., 2022).

After the sales projects are selected, the EEx is obliged to conclude the corresponding contracts, respecting the terms of the public call (Art. 38 R.).

It is worth noting that, in line with the understanding on purchases from family farmers under the PAA,

> in exceptional situations in which the public call procedure is impractical, the adoption of other forms of direct contracting will be admitted, respecting all the restrictions established by the pertinent legislation, not being recommended the use of the electronic reverse auction with exclusive quotas for family farmers.
> (AGU, 2021b)

As discussed in this section, the special rules established for the public call process show an effort to adapt it to the characteristics and capacities of family farming and other vulnerable groups, as well as to regional differences and local specificities. This makes it an appropriate tool to give effectiveness to the reservation scheme provided for by PNAE law and, more broadly, pursue multiple social and environmental objectives.

Sustainable public food procurement beyond the PNAE

Through progressive legislative reforms, sustainable public procurement has become an obligation under Brazilian law.[8] In fact, since 2010 the promotion of sustainable national development has been an objective of public procurement (Art. 3 of Law 8.666/1993) and, under the new general Law 14.133/2021, it is both an objective (Art. 11) and a principle (Art. 5°), and is manifest in a number of provisions. According to the authoritative view,

> the organs and entities of the public administration are required to adopt criteria and practices of social and environmental sustainability and accessibility in public procurement, in the phases of planning, selection of supplier,

> contract execution, supervision. . . and the impossibility of their adoption should be justified by the competent authority in the records of the administrative process, with the indication of relevant fact and/or law grounds.
>
> (AGU, 2021b)

Additionally, public procurement procedures must observe the National Micro and Small Enterprise Statute. This establishes a "favourable, differentiated and simplified treatment to micro and small businesses" in the area of public procurement (Brazil, 2006b). This is materialised through instruments such as preferential contracting in the event of a tie (Art. 44), reserved bidding for contracts under a certain value, mandatory subcontracting in service contracts, and, in supplies contracts, a reserved quota of the subject matter (Art. 48, I, II, III) (see also Art. 4 Law 14.133/2021; Arts. 36, 43, §7° Decree 10.024/2019).

Finally, of particular interest are the gender-related procurement requirements introduced in Law 14.133/2021 by Decree 11.430/2023). This included as a tie-breaker criterion "the development, by the bidder, of equity measures between women and men in the workplace" and the possibility of "requiring in the tender notice that a minimum percentage of the workforce responsible for executing the contract is composed of women who are victims of domestic violence" (respectively, Art. 60, III and Art. 25, §9, I Law 14.133/2021).

While it is not the aim of this study to analyse in detail the general provisions dealing with sustainability in public procurement, it has been shown that, both within and outside the PNAE regulatory framework, there are various strategies for incorporating sustainability considerations into public food procurement.

Conclusions

In Brazil, the National School Feeding Programme stands as a cross-cutting public policy oriented to (i) producers – supporting small-scale agriculture, the development and diversification of agroecological food production, access to markets and income generation for rural communities and vulnerable groups; (ii) markets – aiming at fair prices, shorter supply chains and efficient sales channels; (iii) consumers – improving food security, health and nutrition and encouraging healthy and sustainable eating habits.

The progressive construction of the PNAE legal and regulatory framework for public procurement has been instrumental for harmonising the rights to health, education, environment and social inclusion through school food purchase. Mechanisms such as the reservation scheme through a waiver of bidding, quality-based competition, mandatory food requirements and evaluation criteria prove crucial "to widen the range of development objectives that public entities can reach through public procurement, in a manner that is highly food-specific" (De Schutter et al., 2021). Recent positive reforms have shown a concern to adapt public food demand to the supply capacity of the market (e.g. preferential selection criteria) and also the recent change to provide greater transparency and better monitoring of the policy (e.g. the PNAE card).

In addition to the PNAE, a number of sustainable procurement provisions and mechanisms contained in the general public procurement legislation can also be used for food purchases.

Nonetheless, the wide uptake of sustainable public food procurement in Brazil still faces significant challenges. First, it is important to adjust PNAE funding for inflation and prioritise the areas with lower local revenue and the most socio-economically vulnerable students, given that almost half of the Brazilian municipalities rely entirely (or almost entirely) on PNAE resources to fund school feeding (CMAP, 2020). Additional challenges concern the adaptation of public administration practices to the reality of family farming. A fair purchase price and competition limited to quality and other sustainability aspects must be guaranteed, as well as the purchase of agreed products, timely payments, reasonable flexibility in the documentation required and in the application of penalties during contract performance (CMAP, 2020).

Municipalities' adhesion to the reservation scheme has increased over the years. In 2017 half of all Brazilian municipalities met or exceeded the 30% quota and 86% of all municipalities purchased, to some extent, food from family farms (CMAP, 2020). Structural barriers still hinder the inclusion of family farmers in public procurement. This requires rural development measures to provide technical assistance and training to farmers, investments in the productive infrastructure, and adapting public food demand to local productive capacity, including through collaboration schemes between nutritionists and rural development agencies. There are also mechanisms for encouraging municipalities to observe the family farm purchasing rule (CMAP, 2020; Soares et al., 2021).

It is relevant to mention the "entrepreneurial mayor award" of the Brazilian agency providing support to SMEs (SEBRAE), which rewards projects to expand municipal procurement from local and regional urban and rural small businesses. This includes supporting the adaptation of bidding processes, building businesses' and farmers' capacity to supply to the public administration and training municipal public servants to prioritise the procurement of products and services from these suppliers (SEBRAE, 2021).

The lack of training for suppliers and public administration officials, as well as failures to disseminate information and ensure transparency, are major challenges commonly pointed out in relation to sustainable public procurement. This is reflected most seriously in the context of food procurement, where, despite very detailed regulations, the lack of complete and simplified guidance material for the different players involved hampers effective access to these public policies.

The use of public procurement for food systems transformation is gaining momentum around the world, and calls for closer attention to the connections between the legislative frameworks for public procurement and sustainable development. With this chapter, we hope to have contributed to broadening the discussion on the subject of sustainable public food procurement in Brazil and internationally.

Table 7.1 Abbreviations with Portuguese Wording and English Translation

Abbreviation	*Original title*	*Translation*
AGU	Advocacia-Geral da União	Attorney General of the Union
CAE	Conselho de Alimentação Escolar	School Meals Council
CFN	Conselho Federal de Nutricionistas	Federal Council of Nutritionists
CGU	Controladoria Geral da União	Office of the Comptroller General
CMAP	Conselho de Monitoramento e Avaliação de Políticas Públicas	Public Policy Monitoring and Evaluation Council
COEFA	Coordenação de Execução Financeira da Alimentação	Coordination of the Financial Execution of Food
COSAN	Coordenação de Segurança Alimentar e Nutricional	Coordination of Food and Nutritional Security
CRFB	Constituição da República Federativa do Brasil	Federal Constitution of Republic of Brazil
EEx	Entidade Executora	Executing Entity
FNDE	Fundo Nacional de Desenvolvimento da Educação	National Fund for the Development of Education
IBGE	Instituto Brasileiro de Geografia e Estatística	Brazilian Geography & Statistical Institute
MPF	Ministério público federal	Federal Prosecution Service
PAA	Programa de Aquisição de Alimentos	Food Purchase Programme
PNAE	Programa Nacional de Alimentação Escolar	National School Feeding Programme
PRONAF	Programa Nacional de Fortalecimento da Agricultura Familiar	National Program to Support Family Farming
RT	Nutricionista Responsável Técnico	Responsible Nutritionist
SEBRAE	Serviço Brasileiro de Apoio às Micro e Pequenas Empresas	Brazilian Micro and Small Business Support Service
TCU	Tribunal de Contas da União	Federal Audit Court
UEx	Unidade Executora	Excuting Unit

Notes

1 The author is an early-stage researcher at the SAPIENS Network. This project has received funding from the European Union's Horizon 2020 research and innovation programme under the Marie Skłodowska-Curie grant agreement No 956696. The author would also like to thank the following experts for their valuable research inputs and feedback: Letícia Campos Baird (Bahia State Public Prosecutor), Felipe Jabali Marques and Anna Beatriz Savioli (Brazilian Lawyers and PhD students) and Felippe Vilaça Loureiro Santos (ESR fellow at the SAPIENS Network and PhD student).

2 The PAA's broad scope, consisting of several operating modalities with their own goals and rules, as well as its recent extinction and reintroduction by the Bolsonaro and Lula governments respectively, may justify its exclusion from the scope of this chapter. However, there is no doubt that with almost twenty years of history, the PAA has been a major advance for sustainability in Brazil and deserves special attention in public food procurement studies. See, among others, Souza de Almeida et al., 2020 (cited).

3 In Brazil, the executive branch is divided into three levels: federal, state and municipal, headed by the President at the federal level, governors at state level and mayors at municipal level.
4 At the beginning of 2023, after six years without correction, per capita funding was increased by 28–39% and the total programme's budget reached R$5.5 billion (FNDE, 2023). At the time of writing (September 2023), this corresponded to approx. €1.03 billion and £896 million.
5 This is composed of representatives of government, education professionals and students, parents and civil society organisations.
6 For instance, all the applicants must hold a PRONAF Eligibility Declaration (DAP), certifying compliance with the requirements established by Law 11.326/2006 to classify as a 'family farmer' or 'rural family entrepreneur. An adaptation of the rules on DAP duration – increasing its validity from two to five years – was advocated to reduce potential administrative obstacles to farmers' participation in public calls (CMAP, 2020). Additionally, the applicants must submit a declaration that the food to be supplied is of their production (Art. 36 R.).
7 In September 2023, R$40.000 were equivalent to approx. £6.500 and €7.500.
8 On the mandatory nature of sustainable public procurement in Brazil see, for all, the opinion of the National Sustainability Chamber (AGU, 2021a), and the National Guide for Sustainable Contracting by General Counsel of the Union (AGU, 2021b).

Bibliography

Bravo, H., Sotomayor, O., & Mulder, N. (2022). *Programas de compras públicas a los agricultores familiares: ¿un nuevo canal de ventas para el comercio justo?* [Public procurement programmes from family farmers: a new sales channel for fair trade?]. (LC/TS.2022/130). Santiago: Comisión Económica para América Latina y el Caribe (CEPAL). Retrieved from https://repositorio.cepal.org/bitstream/handle/11362/48109/1/S2200458_es.pdf

Carvalho Filho, J. D. S. (2019). *Manual de Direito Administrativo* [Administrative Law Handbook], 33rd ed. São Paulo: Atlas. Retrieved from www.academia.edu/61549686/Manual_de_Direito_Administrativo_33a_edi%C3%A7%C3%A3o_Jos%C3%A9_dos_Santos_Carvalho_Filho_2019

De Schutter, O., Quinot, G., & Swensson, L. F. J. (2021). Public food procurement as a development tool: the role of the regulatory framework. In: *Public food procurement for sustainable food systems and healthy diets* – Volume 1. Rome: FAO, Alliance of Bioversity International and CIAT and Editora da UFRGS. doi: 10.4060/cb7960en

FAO, Alliance of Bioversity International and CIAT and Editora da UFRGS. (2021). *Public food procurement for sustainable food systems and healthy diets* – Volumes 1 and 2. Rome: FAO, Alliance of Bioversity International and CIAT and Editora da UFRGS. doi: 10.4060/cb7960en; doi: 10.4060/cb7969en

Kluczkovski, A., Menezes, C. A., da Silva, J. T., Bastos, L., Lait, R., Cook, J., Cruz, B., Cerqueira, B., Lago, R. M. R. S., & Gomes, A. N. (2022). An environmental and nutritional evaluation of school food menus in Bahia, Brazil that contribute to local public policy to promote sustainability. *Nutrients*, *14*(7), 1519. doi: 10.3390/nu14071519

Luzzi De Campos, A., & Goldfarb, C. (2021). *Desafios Para O Abastecimento E Soberania Alimentar No Brasil* [Challenges for food supply and sovereignty in Brazil]. Friedrich-Ebert-Stiftung. Retrieved from https://library.fes.de/pdf-files/bueros/brasilien/18104-20210928.pdf

Marques, F. J. (2020). *O protagonismo dos movimentos sociais na conquista da emancipaçaõ social: uma análise da relação entre a cooperativa ACAMPRA – MST e o PNAE em Uberlândia/MG* [The leading role of social movements in achieving social emancipation: an analysis of the relationship between the cooperative ACAMPRA – MST and the PNAE in Uberlândia/MG]. [Undergraduate dissertation in Law. Universidade Federal de Uberlândia]. Retrieved from https://repositorio.ufu.br/handle/123456789/30785

Nehring, R., Miranda, A., & Howe, A. (2017). Making the case for institutional demand: supporting smallholders through procurement and food assistance programmes. *Global Food Security*, *12*, 96–102. doi: 10.1016/j.gfs.2016.09.003.

OECD/FAO. (2022). *OECD-FAO agricultural outlook 2022–2031*. Paris: OECD Publishing. doi: 10.1787/f1b0b29c-en

Rezende, L. T. de, Gottschall, L. M., da Silva Sampaio, K. P., & de Freitas Castro, S. F. (2022). Avanços da legislação do programa nacional da alimentação escolar [Advances in the legislation of the national school feeding programme]. *Cadernos Do FNDE*, *3*(6), 11–33. doi: 10.29327/2138137.3.6-2

Schottz, V., & Schmitt, C. J. (2021). The role of civil society in the construction of the regulatory framework and implementation mechanisms for inclusive public food procurement: the case of the Brazilian national school feeding programme. In *Public food procurement for sustainable food systems and healthy diets* – Volume 2. Rome: FAO, Alliance of Bioversity International and CIAT and Editora da UFRGS. doi: 10.4060/cb7969en

SEBRAE. (2021). *Regulamento Da XI Edição do Prêmio Sebrae Prefeito Empreendedor* [Rules for the 11th edition of the Sebrae mayor entrepreneur award]. Retrieved 28 May, 2023, from www.prefeitoempreendedor.sebrae.com.br/regulamento/

Soares, P., Martinelli, S. S., Davó-Blanes, M. C., Fabri, R. K., Clemente-Gómez, V., & Cavalli, S. B. (2021). Government policy for the procurement of food from local family farming in Brazilian public institutions. *Foods*, *10*(7), 1604. doi: 10.3390/foods10071604

Souza de Almeida, A. F., Perin, G., Mariana, A. P., & Sambuichi, R. H. (2020). *O Programa de Aquisição de Alimentos (PAA): um caso de Big Push Ambiental para a sustentabilidade no Brasil* [The food acquisition programme (PAA): a case of big environmental push for sustainability in Brazil]. Escritório no Brasil da Comissão Econômica para a América Latina e o Caribe (CEPAL). Retrieved from https://archivo.cepal.org/pdfs/bigpushambiental/Caso92-OProgramadeAquisicaodeAlimentos.pdf

Swensson, L. F. J. (2019). Institutional food procurement programmes and producer organizations: catalysts for the transformation of small-scale producers' food supply systems. *Redes*, *24*(1), 30–44. doi: 10.17058/redes.v24i1.13041

Swensson, L. F. J., & Tartanac, F. (2020). Public food procurement for sustainable diets and food systems: the role of the regulatory framework. *Global Food Security*, *25*, 100366. doi: 10.1016/j.gfs.2020.100366

GOVERNMENT SOURCES AND LAWS (in chronological order – from most recent onwards)

The two most important government documents are PNAE Law 11.947 (2009) referred to in the text as L and FNDE (2020) referred to in the text as R. Other relevant legal sources are the general public procurement laws 8.666 (1993) and 14.133 (2021).

AGU. (2021a). *Parecer [Opinion] n. 01/2021/CNS/CGU/AGU*. Retrieved from www.gov.br/agu/pt-br/composicao/cgu/cgu/modelos/licitacoesecontratos/copy_of_PARECER01.2021CNS.pdf

AGU. (2021b). *Guia Nacional de Contratações Sustentáveis* [National Guide for Sustainable Contracting]. 4ª ed. Brasília: *AGU*. Retrieved from www.gov.br/agu/pt-br/comunicacao/noticias/AGUGuiaNacionaldeContrataesSustentveis4edio.pdf

AGU. (2023). *Parecer Referencial [Reference Opinion] n. 0001/2023/COORD/E-CJU/AQUISIÇOES/CGU/AGU*. Retrieved from https://sapiens.agu.gov.br/valida_publico?id=1067943550

Brazil. (1988). *Constituição da Republica Federativa do Brasil de 1988* [Constitution of the Federative Republic of Brazil of 1988]. Retrieved from www.planalto.gov.br/ccivil_03/constituicao/constituicao.htm

Brazil. (1993). *Lei [Law] nº 8.666, de 21 de Junho de 1993 – Regulamenta o art. 37, inciso XXI, da Constituição Federal, institui normas para licitações e contratos da Administração Pública e dá outras providências* [Regulates art. 37, item XXI, of the Federal Constitution, establishes rules for public tenders and contracts and provides other measures]. Retrieved from www.planalto.gov.br/ccivil_03/leis/l8666cons.htm

Brazil. (2006a). *Lei [Law] nº 11.326, de 24 de Julho de 2006 – Estabelece as diretrizes para a formulação da Política Nacional da Agricultura Familiar e Empreendimentos Familiares Rurais* [Establishes the guidelines for the formulation of the national policy for family farming and rural family enterprises]. Retrieved from www.planalto.gov.br/ccivil_03/_ato2004-2006/2006/lei/l11326.htm#:~:text=LEI%20N%C2%BA%2011.326%2C%20DE%2024%20DE%20JULHO%20DE%202006.&text=Estabelece%20as%20diretrizes%20para%20a,Familiar%20e%20Empreendimentos%20Familiares%20Rurais.

Brazil. (2006b). *Lei Complementar [Complementary Law] nº 123, de 14 de Dezembro de 2006 – Institui o Estatuto Nacional da Microempresa e da Empresa de Pequeno Porte* [Establishes the National Statute for Microenterprises and Small Businesses]. Retrieved from www.planalto.gov.br/ccivil_03/LEIS/LCP/Lcp123.htm

Brazil. (2009). *Lei [Law] nº 11.947, de 16 de Junho de 2009 – Dispõe sobre o atendimento da alimentação escolar [. . .]* [Provides for school nutrition [. . .]]. Retrieved from www.planalto.gov.br/ccivil_03/_ato2007-2010/2009/lei/l11947.htm

Brazil. (2014). *Lei [Law] nº 12.982, de 28 de Maio de 2014 – Altera a Lei nº 11.947, de 16 de junho de 2009, para determinar o provimento de alimentação escolar adequada aos alunos portadores de estado ou de condição de saúde específica* [Amends Law n° 11.947, of June 16, 2009, to determine the provision of adequate school meals for students with specific health conditions]. Retrieved from www.planalto.gov.br/ccivil_03/_ato2011-2014/2014/lei/l12982.htm

Brazil. (2019). *Decreto [Decree] nº 10.024, de 20 de Setembro de 2019 – Regulamenta a licitação, na modalidade pregão, na forma eletrônica, para a aquisição de bens e a contratação de serviços comuns* [Regulates electronic auctions for the purchase of goods and the contracting of common services]. Retrieved from www.planalto.gov.br/ccivil_03/_ato2019-2022/2019/decreto/d10024.htm

Brazil. (2021). *Lei [Law] nº 14.133, de 1º de Abril de 2021 – Lei de Licitações e Contratos Administrativos* [Public tenders and contracts law]. Retrieved from www.planalto.gov.br/ccivil_03/_ato2019-2022/2021/lei/L14133.htm

Brazil. (2023a). *Decreto [Decree] nº 11.430, de 8 de Março de 2023 – Regulamenta a Lei nº 14.133, de 1º de abril de 2021, para dispor sobre a exigência, em contratações públicas, de percentual mínimo de mão de obra constituída por mulheres vítimas de violência doméstica e sobre a utilização do desenvolvimento, pelo licitante, de ações de equidade entre mulheres e homens no ambiente de trabalho como critério de desempate em licitações, no âmbito da administração pública federal direta, autárquica e fundaciona*. [Regulates Law n° 14.133, of April 1, 2021, to provide for the requirement, in public contracts, of a minimum percentage of workforce made up of women victims of domestic violence and on the use of the development, by the bidder, of equity actions between women and men in the work environment as a tiebreaker criterion in bids, within the scope of the direct, autarchic and foundational federal public administration]. Retrieved from www.planalto.gov.br/ccivil_03/_ato2023-2026/2023/decreto/D11430.htm

Brazil. (2023b). *Lei [Law] nº 14.628, de 20 de julho de 2023 – Institui o Programa de Aquisição de Alimentos (PAA)* [. . .]. [Establishes the Food Acquisition Program (PAA) [. . .]]. Retrieved from www.planalto.gov.br/ccivil_03/_ato2023-2026/2023/lei/l14628.htm

CFN. (2010). *Resoluçao [Resolution] nº 465, de 23 de Agosto de 2010 – Dispõe sobre as atribuições do Nutricionista, estabelece parâmetros numéricos mínimos de referência no âmbito do Programa de Alimentação Escolar (PAE) e dá outras providências* [Provides for the duties of the Nutritionist, establishes minimum quantitative parameters for the National School Feeding Program (PNAE) and other provisions]. Retrieved from http://sisnormas.cfn.org.br:8081/viewPage.html?id=465

CMAP. (2020). *Relatório de Avaliação Programa Nacional de Alimentação Escolar (PNAE)* [Evaluation report of the National School Feeding Programme – PNAE]. Retrievd from www.gov.br/economia/pt-br/acesso-a-informacao/participacao-social/conselhos-e-orgaos-colegiados/cmap/politicas/2020/gastos-diretos/relatorio_avaliacao-cmag-2020-pnae.pdf

COEFA. (2020). *Nota Técnica nº 1904244/2020/COEFA/CGPAE/DIRAE – Alterações constantes da Resolução CD/FNDE nº 6, de 8 de Maio de 2020, relativas aos aspectos relacionados aos recursos financeiros federais do programa nacional de alimentação escolar (pnae) e sua movimentação* [Amendments in Resolution CD/FNDE n° 6, of May 8, 2020, concerning federal financial resources of the National School Feeding Programme (PNAE) and their management]. In *Caderno de legislação 2022*. FNDE. (2022). Retrieved from www.gov.br/fnde/pt-br/acesso-a-informacao/acoes-e-programas/programas/pnae/manuais-e-cartilhas/CADERNODELEGISLAO2022atualizadaltimaverso.pdf

COSAN. (2016). *Nota Técnica [Technical Note] nº 5007/2016/COSAN/CGPAE/DIRAE – Especificação de gêneros alimentícios para o programa nacional de alimentação escolar – PNAE* [Specification of foodstuffs for the National School Feeding Programme]. Retrieved from www.gov.br/fnde/pt-br/acesso-a-informacao/acoes-e-programas/programas/pnae/media-pnae/notas-tecnicas/2016/Notatcnica_5007_2016.pdf

COSAN. (2020a). *Nota Técnica [Technical Note] nº 1879810/2020/COSAN/CGPAE/DIRAE – Alterações dos aspectos de alimentação e nutrição e de segurança alimentar e nutricional da resolução CD/FNDE nº 6, de 8 de Maio de 2020* [Changes to the food and nutrition and food and nutritional security aspects of Resolution CD/FNDE nº 6, of May 8, 2020]. Retrieved from www.fnde.gov.br/phocadownload/programas/alimentacao_escolar/2020/Nota%20Tecnica_nutrio_PNAE.pdf

COSAN. (2020b). *Nota Técnica [Technical Note] nº* 1894673/2020/COSAN/CGPAE/DIRAE – *Atualização das recomendações acerca da alimentação vegetariana no programa nacional de alimentação escolar – PNAE* [Updating the recommendations on vegetarian nutrition in the National School Feeding Programme – PNAE]. Retrieved from www.gov.br/fnde/pt-br/acesso-a-informacao/acoes-e-programas/programas/pnae/media-pnae/notas-tecnicas/2020/NotaTcnican1894673AtualizaodasrecomendaesacercadaalimentaovegetariananoProgramaNacionaldeAlimentaoEscolarPNAE.pdf

COSAN. (2020c). *Nota Técnica [Technical Note] nº 1897361/2020/DIDAF/COSAN/CGPAE/DIRAE – Alterações dos aspectos da agricultura familiar da resolução CD/FNDE n° 6, de 8 de maio de 2020* [Changes to the family farming aspects of Resolution CD/FNDE n° 6, of May 8, 2020]. Retrieved from www.gov.br/fnde/pt-br/acesso-a-informacao/acoes-e-programas/programas/pnae/media-pnae/notas-tecnicas/2020/SEIFNDE1897361NotaTecnica.pdf

FNDE. (2020). *Resolução [Resolution] nº 6, de 08 de Maio de 2020 – Dispõe sobre o atendimento da alimentação escolar aos alunos da educação básica no âmbito do Programa Nacional de Alimentação Escolar – PNAE* [On school feeding provision to students of basic education under the National School Feeding Programme – PNAE]. Retrieved from www.gov.br/fnde/pt-br/acesso-a-informacao/legislacao/resolucoes/2020/resolucao-no-6-de-08-de-maio-de-2020/view

FNDE. (2023, 10 March). *Governo Federal reajusta valores da alimentação escolar* [Federal Government readjusts school feeding budget]. Retrieved 29 May, 2023, from www.gov.br/fnde/pt-br/assuntos/noticias/governo-federal-reajusta-valores-da-alimentacao-escolar

IBGE. (2019). *Censo agropecuário: resultados definitivos 2017* [Agricultural Census: definitive results 2017]. Rio de Janeiro: Instituto Brasileiro de Geografia e Estatística. Retrieved from https://biblioteca.ibge.gov.br/visualizacao/periodicos/3096/agro_2017_resultados_definitivos.pdf

MPF. (2020). *Nota Técnica [Technical Note] nº 3/2020/6ªCCR/MPF – Serviços de inspeção sanitária incidentes sobre a comercialização e consumo de alimentos produzidos pelos povos e comunidades tradicionais* [Sanitary inspection services for the marketing and consumption of food produced by traditional peoples and communities]. Retrieved from www.mpf.mp.br/atuacao-tematica/ccr6/catrapovosbrasil/documentos-e-publicacoes/nota_tecnica_geral_-_3-2020_6ccr_-_alimentos_tradicionais.pdf

8 Sustainable food procurement and universities

A participative and eco-innovative tender for the vending machines at the University of Turin

Nadia Tecco and Egidio Dansero

Introduction

Public Food Procurement is one of the most powerful tools available to public authorities in the context of "second-generation" environmental policies to drive innovation from the demand side. Universities can exercise leverage towards orienting the food system towards sustainable food consumption. Their example can influence the behaviour of other organisations within the city and the region.

The chapter presents some insights about the university's role in sustainable food procurement derived from experience of a tender for food and drink vending concession services at the University of Turin carried out within the framework of the European SPP Regions project.[1] Distinctive features of this case history were the shift from a product logic to a "supply chain" logic in the stakeholder engagement process that led to the preparation of the tender, as well as the integration of green and circular criteria in the contract model.[2]

The potential of public food procurement to work on the link between food systems, nutrition, and health and to make a player like the university a game changer for food system transformation is therefore confirmed. However, as emerges from the case study, there are certain necessary upstream and downstream conditions. Upstream, objectives and guidelines established by the national dietary and nutritional guidelines need to be translated into criteria reachable by the market in terms of technical feasibility and price for the final consumer. Moreover, the introduction of healthy foods in a vending machine should be promoted by greater accessibility (in terms of sales price and visibility) compared to unhealthy food, alongside with the development of ad hoc educational campaigns for the target community. Downstream, there needs to be an adequate monitoring process for the entire execution of the contract.

Universities and the challenge of sustainability

Increasingly, universities are being called upon to respond to the challenges of contemporary society, no longer by means of research that can only be used by academia, but by acting directly within their everyday operations. This shift is well expressed by the title of a text by Judith Rodin (2007) on universities in urban

DOI: 10.4324/9781003393023-10

contexts: "Out of the ivory tower and into the streets". For a discussion of how universities have faced the challenge of sustainability see Soini et al. (2018). The challenges facing the civic university are discussed by Goddard et al., 2016. Chatterton (2000) and Chatterton (2010) consider the cultural and economic roles of universities within the urban communities where they are located.

Greening Universities is a theme that has long seen a plurality of both initiatives and theoretical reflections, systematising empirical experiences with an intense debate on universities as promoters of sustainability. The challenge posed by the 2030 Agenda can indeed be seen in two perspectives. On the one hand, how can the sustainable development goals (SDGs) help universities and conversely how can universities help the process of achieving the SDGs?

The desire to achieve SDGs involves universities in their different missions, research, education, third mission and public engagement, with sustainability as a challenge that is relevant to all of these and that must be translated into choices that affect both the governance and operational activities of universities. These choices also affect their important external leadership role in orienting and influencing public debate within the wider community.

Universities frequently refer to their third mission. After research and education, universities have an important economic and social role and make a significant contribution to the communities and territories within which they operate (Compagnucci & Spigarelli, 2020).

Within the third mission, universities are called upon to develop targeted actions of networking with local authorities, stakeholder engagement and active citizenship involvement. The development of actions aimed at improving the University's overall environmental performance and represents a permanent laboratory of active citizenship that requires the involvement and commitment of the entire university community, from the governing bodies to the student community, from lecturers to technical-administrative staff.

Universities are committed in various ways to the exchange of good practices and international comparison. This may include participation in networks such as International Sustainable Campus Network (ISCN), UN Sustainable Development Solutions Network (SDSN) and in the Italian context the RUS (Universities Network for Sustainable Development). Use may also be made of sustainability toolkits devised for universities by SDSN and the United Nation Environment Programme (UNEP). Universities may sign up to charters and commitments such as the Talloires declaration by University Leaders for a Sustainable Future (http://ulsf.org/talloires-declaration/).

Universities may participate in specialised ranking initiatives such as Green Metric (https://greenmetric.ui.ac.id/). They may also be involved in more general initiatives which have widened their scope to the themes of sustainability and social impact (www.topuniversities.com/).

Universities policies affect multiple areas of operation: energy, resource and waste management, food, mobility, logistics and administrative services.

Public procurement represents the main lever for reorienting universities as structures and organisations, touching on all areas of operation and playing an important role vis-à-vis society and the market, especially if this is done in a coordinated and networked action with other universities and public bodies.

Food is an important point of intervention for universities, considering university canteens and cafeterias, catering and vending machines. Sustainable food procurement and other initiatives can make an active contribution as laboratories of good practices for the promotion of healthy, ethical and environmentally sustainable food (Doherty et al., 2011).

Universities and sustainable food procurement: How and why engage with vending machines?

By intervening, for instance, on tender specifications for university canteens or vending machines or even through food production activities within university experimental farms, universities can become vectors of culture, awareness and innovation, thus playing a role that generates multiple impacts and values in relation to different stakeholders (Leal Filho et al., 2022; Mohammed et al., 2022).

The application of green public procurement therefore appears to be one of the most suitable actions demonstrating the willingness of universities to engage in the pursuit of the sustainable development goals as defined by the 2030 Agenda, which explicitly calls for the promotion of sustainable public procurement practices, in accordance with national policies and priorities as one of the targets of Goal 12 "Achieving sustainable patterns of production and consumption" (Sirsi, 2019). Public Food Procurement becomes a powerful instrument of second-generation economic policy, not only because it can reduce the direct environmental impact of public authorities and represent an example of environmental awareness, but also because of the possibility of fostering the integration of environmental considerations into other policies (such as transport, energy, waste reduction and management) and of exerting pressure on the market to offer goods, services and projects with lower environmental impact, promoting the development of technologies and supply chains and the growth of green companies (Swensson & Tartanac, 2020). For all these reasons the spread of green procurement, can be considered a fundamental tool for solving different problems at the same time and a key factor in a transformative perspective (Swensson et al., 2021). Dealing with the vending machine service within universities and other public administration entities involves pursuing food, nutrition and environmental education, with the aim of consolidating healthy habits and increasing environmental protection practices in a comprehensive manner.

The vending machine sector, excluding the parenthesis of the pandemic period, has been characterised by a strong market expansion in recent years. With the sale of 99 million food and beverage products every day in Europe and an average of 185 Europeans per vending machine (European Vending and Coffee Service, 2022), it poses significant challenges in terms of environmental sustainability in relation to food, nutritional intake, production and waste management, energy consumption, procurement and logistics.

For some years now, the environmental performance of vending machines, given their pervasiveness in public and private spaces and the volumes of food and drink sales and waste produced (especially single-use plastic), has attracted public attention. Consumers are increasingly sensitive to environmental and health issues relating to food and drink vending machines.

Although there have been small improvements, the university vending machine sector is still far from being a model of sustainability. In the face of consumer demand, some operators have begun, albeit timidly, to experiment with actions to adapt to a market that is turning towards a more marked attention to respect for the environment, with respect to the food offer, management of the service and the efficiency of the vending machines as such.

These are changes in consumption styles and environmental perception/sensitisation to which the vending industry is trying to adapt and incorporate innovations into its performance (Bertossi et al., 2022).

In 2020, the then Ministry of the Environment organised a series of consultation meetings with the intention of defining Minimum Environmental Criteria (Criteri Ambientali Minimi, CAM) for vending machines with the aim of promoting eco-design criteria, specific environmental technologies and better management solutions from an energy, environmental and ethical-social point of view. Such consultation meetings are commonly referred to as "working tables" [*Tavoli di lavoro*]. The "working table" was attended by public organisations which purchase vending machines and trade associations representing vending machine suppliers. There followed more than two years of discussion. A proposal document was drawn up and drafted and redrafted. There was a complicated and tough confrontation between polarised positions that were difficult to reconcile. Finally, the process was interrupted due to the lack of an agreement between the parties, effectively blocking the adoption of Minimum Environmental Criteria. The Vending Trade Associations rejected the draft proposal, complaining that the proposed rules and criteria were inconsistent with the very nature of the vending service.

Among those who supported the proposal was a representative of the Network of Universities for Sustainable Development in which the University of Turin participates. The proposal favoured an improvement in the sustainability of the sector, reducing environmental impact and promoting public health. The aspiration was to protect the public and collective interest and to use CAM to make the public procurement contract not only a tool for the acquisition of goods and services, but an economic policy tool effective not only in reducing environmental impacts, also encouraging more sustainable, "circular" production and consumption models and in promoting green employment. This has been advocated by a recent Council of State ruling (Italian Republic, 2022; Iurascu, 2023).

The outcome of this consultation confirms the difficulty, at times bordering on short-sightedness displayed by suppliers with taking systematic action to improve the environmental performance and nutritional profile of their products. They were slow to respond both to policy guidelines at European level (such as Farm to Fork Strategy) and to the needs of consumers who increasingly preferred products consistent with a healthy and sustainable diet.

Developing the vending machine tender

The establishment of UniToGO, the Green Office of the University of Turin, in May 2016, signalled the university's increased commitment to define and implement environmental actions. Priority areas included food procurement, energy, waste, mobility and climate change. The goal was to "increase the social, economic and environmental responsibility of the University of Turin as stated in the University Strategic Plan" (Università degli Studi di Torino, 2016; 2021).

Starting in spring 2017, the Food working group and the Green Public Procurement group, in close cooperation with the other UniToGO working groups (Waste, Mobility and Energy), coordinated the governance process that led to the definition and drafting of an eco-innovative set of specifications for the concession of the installation and management of food and beverage vending machines at university sites. The governance process covered both the contextual aspects concerning the requirements of the tendering process for vending services and the procedural aspects of how the different actors within the university were brought together and consulted for the definition of the tender.

The Green Office's contribution took the form of defining technical specifications and eco-innovative criteria to improve environmental sustainability of the service, while guaranteeing good quality products and adequate affordability for 80,000 students and over 5,000 research and technical-administrative staff.

The entire process followed in general terms the principles of eco-innovation as defined by the European Commission in its communication (European Commission, 2011), together with the principles of affordability, widespread presence and accessibility to a healthy diet as defined by the international debate and in particular by World Health Organization (2004) and Agriculture Ministry (2003). The overall goals pursued by the vending machine tender are summed up in Figure 8.1.

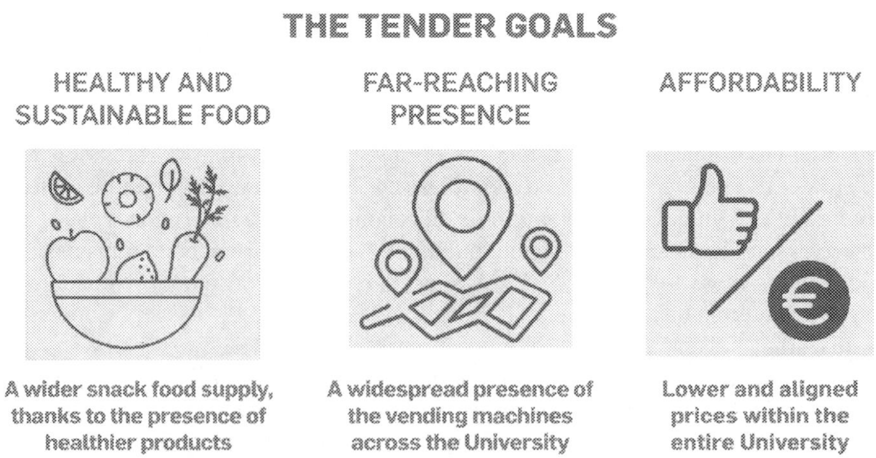

Figure 8.1 The overall goals pursued by the vending machine tender

Source: Nadia Tecco

The drafting of the eco-innovative vending specifications was included as a pilot study within the H2020 SPP Regions project, a network of European institutions and municipalities to foster the adoption of eco-innovative solutions and the sharing of experiences in the field of sustainable public procurement (SPP) and public innovation procurement (PPI), including the topic of food and beverage services.

Through its participation in this project, UniTO seized the opportunity to trigger a process of reorganisation of the University's food and beverage vending service, which would at the same time improve its environmental sustainability requirements and favour alternative forms of procurement.

The partnership of the SPP Regions project has itself developed from the collaboration that UniTO, and specifically the Green Public Procurement working group, has developed since 2014 with the Provincial Network of Green Public Procurement (*Rete Provinciale Acquisti Pubblici Ecologici – APE*). This is an initiative set up by the Metropolitan City of Turin, to encourage purchasing and environment offices of the provincial authorities to integrate environmental criteria into procurement procedures. It is supported by the Regional Agency for Environmental Protection (*Agenzia Regionale di Protezione Ambientale* – ARPA Piedmont).

Multilevel stakeholder engagement

For the drafting of the tender and its technical contents, a stakeholder engagement process lasting approximately six months was carried out, involving actors both internal and external to the university community. The participation of the internal community was solicited either as a future user of the service or as a bearer of specific technical expertise in defining the tender criteria for a complex supply chain. The involvement of external subjects took the form of dialogue and information sharing with operators in the sector to consider innovation opportunities to improve quality (nutritional and product range offerings) and reduce environmental impact. The objective was co-creation of value in order to increase the acceptability of the changes made to the service. The starting point was a needs analysis and the definition of the status quo in conjunction with the initiation of a community engagement process. Between April and June 2017, a census was carried out on the location/number/type/model of vending machines in different buildings and the status quo of contracts in cooperation with the different departments and administrative structures concerned. UniToGO carried out energy consumption measurements for the different types and models of vending machines (hot drinks, cold drinks, snacks) to obtain information on overall energy consumption.

A total of 279 vending machines were identified, located in 60 university buildings (offices, libraries, classrooms, laboratories, departments) throughout the city. The census showed a varied picture in terms of suppliers (15), contractual conditions, types of distributors, goods and services offered, prices for users. The preliminary analysis therefore revealed a clear need to regulate the automated vending service within a common frame of reference and a wide margin for improving the quality and efficiency of the service. In several locations, the results of the census cross-checked with sales data showed that the number of vending machines was excessive in relation to the number of potential users, with a possible margin for downsizing of about 15% on average.

To define the volume of consumption and to estimate the overall economic value of the tender, market volume data from 2017 onwards were collected from the economic operators then present in the University's areas. Alongside this operation, data collection campaigns were also carried out by direct observation of selected vending machines on a sample basis.

In July 2017, the entire university community was invited to answer a short online questionnaire on "what future for vending machines. Take part in a short survey" [*Quale futuro per i distributori automatici. Participa ad un breve sondaggio*] (Figure 8.2). A total of 1,245 responses were collected with a balanced participation of the three components of the UniTO community: students, lecturers and technical and administrative staff. Analysis of the questionnaire revealed a strong interest in the need for eco-innovative solutions to improve sustainability. Users were particularly receptive to increasing energy efficiency.

Users responding to open-ended questions expressed strong interest in healthier food and beverages and food suitable for different dietary needs (diets, allergies, intolerances). Students wanted water dispensers to reduce usage of plastic bottles.

On 24 July 2017, a preliminary market consultation procedure was initiated, pursuant to Article 66 of Legislative Decree No. 50/201613 of the Public Contracts Code, prior to the start of the tender procedure. Economic operators in the sector were invited to participate in the consultation in which they were asked to answer several questions on the technical feasibility of eco-functionality requirements for vending machines and the capacity of the market to offer eco-innovative solutions. The University met with nine economic operators. From the market

Figure 8.2 The poster used to promote participation in the questionnaire through the University's institutional channels and UniToGO's social channels

Source: Nadia Tecco)

responses, it became clear that some suppliers (particularly the larger ones) were moving towards offering a more sustainable vending service. In the case of new technologies or the management of business peculiarities that could have represented a competitive advantage, the responses from companies were vague and general. Respondents may have been reluctant to share information which might help competitors.

Regarding the feasibility of inserting mains water dispensers, the companies were united in emphasising the lack of feasibility of the request unless there was a large increase in the operating budget. Similar opposition arose when companies were asked to replace plastic coffee cups with other materials such as paper or bioplastic.

In the process of drafting the call for tenders, recourse was had to the expertise from various departments and directorates on legal, economic, environmental, energy and food issues connected to the vending service, creating synergies among the classic triad of research staff, technical/administrative staff and student body that characterises universities.

The dialogue with the SPP Regions network was also of great importance, especially with the local partners of the project, Metropolitan City of Turin and ARPA Piedmont. There was also an exchange of information through the Italian University Network for Sustainable Development (*Rete delle Università per lo Sviluppo Sostenibile* – RUS) with other Italian universities which had also published concession tenders for food and beverage vending machine services. The intention was to assess the most recent developments in legislation and consider critical issues encountered in the course of these projects. Networking with these universities helped us to learn about the critical issues which they had encountered in developing similar tenders. Various RUS working groups were helpful – particularly that focusing on food. Valuable assistance was also provided by the Italian Network of Local Food Policies (www.politichelocalicibo.it).

This collaboration made it possible to compare and systematise good practices, to promote proposals and to unblock inertia and apathy, grasping the overall sense of a process towards a necessary paradigm shift that would otherwise risk being missed when considering individual activities and initiatives.

Whole system criteria design to balance and boost sustainability with concrete feasibility

For the achievement of the objectives and the consequent selection of criteria, a benchmarking analysis was carried out in parallel to the participation activities.

The working group Food and GPP of UniToGO analysed the most relevant eco-features in 20 green contracts from universities and public institutions (ministries, local authorities). The aim of this analysis was to find best practices and new starting points to be adapted to the context of the university. The analysis aimed to build upon the Environment Ministry "Environmental Minimum Criteria" (*Criteri ambientali minimi*) for canteens and public food procurement (Gazzetta Ufficiale della Repubblica Italiana, 2011) and the EU Joint Research Group guidelines (Joint Research Centre, 2016).

The working group followed UniToGO's approach to sustainability of "Know, Involve & Change" (Green Office – Università degli Studi di Torino, 2018). It adopted a whole system criteria design approach, considering food, energy, waste, and mobility, (Doherty et al., 2011). The working group agreed a set of minimum and award criteria which were included in the tender in order to push market operators to propose eco-innovative solutions that would deliver a more efficient and sustainable service. With the tender evaluation the working group chose to award 70% of the score to the technical evaluation which considered innovation and sustainability and price only received 30% of the score.

With the ambition of promoting access to healthy products, all snack and beverage machines were required to hold a minimum of eight snacks of a high nutritional standard, selected with the advice from nutritionists. These included fruit and vegetable products as well as sweet and savoury products compatible with vegetarian and vegan eating styles and considering common allergies and food intolerances. These products were required to have no more than specified maximum levels of added sugars, flavour enhancers, polyphosphates, trans-fatty acids or salt, based on recommendations from the World Health Organisation (WHO) and the Italian Society of Human Nutrition (SINU). The eight healthy snacks were to be clearly labelled in such a way as to highlight their high nutritional quality. Provision of fresh and organic products was encouraged.

Regarding water the vending machine operator was required to provide bottled natural mineral water, for which source provenance was requested as a basic requirement. Additional points were awarded if the water originated within a radius of 100 km from Turin. Vending machine operators were also requested to experiment with the supply of water from dispensers connected to the waterpipe network. It was anticipated that this could lead to a minimum saving in consumption of 99,000 bottles per year. This would be cheaper than bottled water and reduce the consumption of single-use plastics in line with Italian universities' previous commitments (Coco, 2019).

On the energy front, the specification opted for a medium-high energy efficiency class (A) according to the standard set by the European Vending Association Energy Management Protocol (EVA EMP). The specification also required fitting of integrated LED lighting systems for all vending machines. Market surveys had shown that energy class A had been widely used in the latest generation of vending machines for at least the previous two years. A bonus score was included for those who offered vending machines in energy class A+ or higher according to EVA EMP 3.0 or the official European energy standard EN50597.

Based on the spatial and organisational distribution of the University buildings, the tender specifications were structured into five lots, coinciding with the current University centres, considering various principles (logistical, functional, size of the specifications, as well as the need to allow access to micro-small and medium-sized enterprises, as required by the Public Contracts Code). In addition to this, to avoid the risk of any collusive agreements between the participants, the tender rules included a constraint of awarding no more than one lot to each company participating in the tender.

The tender worth more than 9 million euros for three years (renewable) received a discrete number of bids (an average of seven out of the planned five lots). The solutions proposed by the operators chosen as service concessionaires were able to adequately meet the requirements and criteria included in the tender.

Previously there was autonomous vending machine management by various departments/areas/schools/libraries. The new arrangement was a centralised management model, in which in each lot there was only one manager in charge, operating the same pricing policy and with practices based on the principles of the tender specification.

The tender was awarded in July 2018, with commencement of the eco-innovative vending service in January 2019.

As the main final outcomes, the food supply of the vending machine in the University has been enriched with the addition of healthy snacks and a balanced nutritional intake, tap water dispensers have been introduced in the university buildings along with an overall improvement of the energy efficiency of the vending machines with improved energy class (A+) and LED lighting. Some amelioration actions for the waste management were also introduced such as the use of plastic compactors and the collection and re-use of coffee grounds.

The decrease in turnover for operators during the pandemic period, caused initially by the closure of the university and then by the reduced attendance due to staff and student facilities for agile work and distance learning, prompted some operators to request variation of some contractual clauses in relation to the presence of fresh/healthy products and pricing, which was further affected by recent inflation.

Conclusions and hindsight reflections

This experience represented an important achievement for the University in terms of process and increasing the level of eco-innovation and reducing the environmental impact of the vending service. The tender covered the three-year period, 2019–2022. The duration of the concession was extended by one year to take into account the period (March 2020–January 2022) when in-person classes were suspended during the COVID-19 pandemic.

The university has initiated further work with the involvement of the Green Office on enhancing sustainability in the other services that affect food consumption in the university, particularly bars and catering.

Intervening with a single set of specifications in a very uneven and diverse situation represented an opportunity to spread principles of environmental sustainability, efficiency and multidimensional accessibility. The vending service can thus represent a vector of sustainability in the food practices of the community in a public institution.

The struggle to reach an agreed list of environmental criteria for the vending sector which took place at the Working Table initiated by the Ministry of the Environment helped people in the university to clarify their thinking as to how to make vending more sustainable.

A process of sharing best practices in the drafting and management of services related to food consumption in universities was initiated within the network RUS with the aim of producing agreed guidelines among universities.

The case study thus testifies how, through the practices of green public procurement of products and services, it is possible to play a key role on the demand side to support the production of goods and services that are preferable from the point of view of environmental impact and act as a driver in the process of directing more sustainable consumption choices.

It should be emphasised, however, that this requires considerable efforts and success is not guaranteed. There is a need for strong institutional support along with the identification of dedicated and competent human resources, which will transform green purchasing and green public procurement into a tool to support an integrated sustainability policy for our university as well as for all other institutions where such an experience could be replicated. We are thinking here specifically to the universities that adhere to the RUS and in particular the Food Working Group and other public and non-public contexts, such as schools of other types and levels, municipalities, provinces, companies that wish to promote and educate on responsible and environmentally sustainable consumption.

The skills required, in addition to procurement expertise, must include the European Commission sustainability competence framework (Bianchi et al., 2022). There are specific skills required in relation to the vending industry (such as applying nutritional criteria to products marketed in vending machines). The procurement officer must seek to overcome the information asymmetry (Baron & Besanko, 1987), that characterises the sector. They must also keep up to date with innovation and legislation affecting the sector – such as the Energy Labelling Regulation (2019), the Single Use Plastic Directive (2019) and the Eco-design Regulation (2019).

Furthermore, downstream, during the execution of the contract, constant monitoring is required, the conditions and tools for which must be already created during the procurement planning stage. There will be a need for procurement management plans, contractor performance reports, approved change requests and periodical checks carried out by the university.

For multi-site and complex organisations such as our university, monitoring could directly involve the university community through the creation of local committees for community-based monitoring (Olen, 2023).

It should be considered how best to communicate activities internally, especially where changes have been implemented that while virtuous could be perceived as a deterioration of service (reduction of sweetened and carbonated beverages, abolition of sale of beverages in plastic bottles, elimination of super processed food).

Beyond basic monitoring and reporting on the application of GPP, universities should also, as research organisations, consider how to quantify the contribution of GPP by measuring its impact in terms of CO_2 emissions, energy consumption, waste and other environmental targets. There will be a need to establish a baseline at the commencement of each new contract so that progress can be measured.

Notes

1 Within SPP Regions, seven European regional municipality networks work together on Sustainable Public Procurement (SPP) and public procurement of innovation (PPI). The regional networks transfer skills and knowledge through their SPP and PPI activities with a focus on SPP tendering in the field of energy use in public buildings, vehicles/transport and in food and catering services. https://iclei-europe.org/projects/?SPP_Regions-Regional_Networks_for_Sustainable_Procurement_&projectID=uUkX8W8r.

2 A preliminary version of the contents of this chapter was published in Tecco, N., Dansero, E., (2019). Le Vending Machine: Un capitolato eco-innovativo per l'Università degli Studi di Torino. [Vending Machines: An Eco-Innovative Tender for the University of Turin]. In Dansero, E., Fassio, F., Sirsi, E., Tecco, N. (eds.), *Le università e il cibo Buone pratiche verso un'alimentazione sostenibile negli atenei* [Universities and food. Good Practices Towards Sustainable Food Universities and Food], Celid, Torino. Retrieved from www.celid.it/scheda-libro?aaref=1341.

References – general

Agriculture Ministry. (2003). Ministero delle Politiche Agricole e Forestali; & Istituto Nazionale di Ricerca per gli Alimenti e la Nutrizione (INRAN). *Linee Guida per una Sana Alimentazione Italiana* [Guidelines for Healthy Italian Diet]. Retrieved from www.salute.gov.it/imgs/C_17_pubblicazioni_652_allegato.pdf

Baron, D. P., & Besanko, D. (1987). Monitoring, moral hazard, asymmetric information, and risk sharing in procurement contracting. *The RAND Journal of Economics*, 509–532.

Bertossi, A., Troiano, S., & Marangon, F. (2022). Where is sustainability? An assessment of vending products. *Rivista di Studi sulla Sostenibilità*, *XII*(1), 155–180. Retrieved from http://digital.casalini.it/10.3280/RISS2022-001010

Bianchi, G., Pisiotis, U., & Cabrera Giraldez, M. (2022). GreenComp the European sustainability competence framework. In Punie, Y., & Bacigalupo, M. editor(s), *EUR 30955 EN*, Publications Office of the European Union, Luxembourg. ISBN 978-92-76-53201-9, doi: 10.2760/13286 JRC128040. Retrieved from https://op.europa.eu/en/publication-detail/-/publication/bc83061d-74ec-11ec-9136–01aa75ed71a1/language-en, https://dx.doi.org/10.2760/13286

Chatterton, P. (2000). The cultural role of universities in the community: Revisiting the university – community debate. *Environment and Planning*, *32*, 165–181.

Chatterton, P. (2010). The student city: An ongoing story of neoliberalism, gentrification, and commodification. *Environment and Planning*, *42*, 509–514.

Coco, A. (2019, 18 March). *Scuola e università diventano Plastic Free* [School and University Become Plastic Free]. Retrieved from https://anteritalia.org/scuola-e-universita-diventano-plastic-free/

Compagnucci, L., & Spigarelli, F. (2020). The third mission of the university: A systematic literature review on potentials and constraints. *Technological Forecasting and Social Change*, *161*, 120284. doi: 10.1016/j.techfore.2020.120284

Doherty, S., Cawood, J., & Dooris, M. (2011). Applying the whole-system settings approach to food within universities. *Perspectives in Public Health*, *131*(5), 217–224. doi: 10.1177/1757913911413344

European Commission. (2011). *Communication from the commission to the European parliament, the council, the European economic and social committee and the committee of the regions – innovation for a sustainable future – the Eco-innovation Action Plan*

(Eco-AP). Bruxelles. Retrieved from https://eur-lex.europa.eu/legal-content/EN/TXT/HTML/?uri=CELEX:52011DC0899

European Vending and Coffee Service. (2022). *Europe the vending market in 2021*. Retrieved from www.vending-europe.eu/activities/market-information/

Gazzetta Ufficiale della Repubblica Italiana. (2011). *Adozione dei criteri minimi ambientali da inserire nei bandi di gara della Pubblica amministrazione per l'acquisto di prodotti e servizi nei settori della ristorazione collettiva e fornitura di derrate alimentari e serramenti esterni* [Adoption of minimum environmental criteria to be included in public administration tenders for the purchase of products and services in the sectors of catering and food supply and external joinery]. n.220 21–09–2011. Retrieved from www.gazzettaufficiale.it/eli/id/2011/09/21/11A12078/sg

Goddard, J., Hazelkorn, E., Kempton, L., & Vallance, P. (2016). *The Civic University: The Policy and Leadership Challenges*, Edward Elgar, Cheltenham.

Green Office – Università degli Studi di Torino. (2018). *Piano d'azione per la sostenibilità ambientale di Ateneo* [Action Plan for the Environmental Sustainability of the University of Turin]. Università degli Studi di Torino. Retrieved from http://politichediateneounito.it/wp-con-tent/uploads/2018/11/UniToGO-Piano-di-Azione 2018–2020.pdf

Italian Republic. (2022). *Consiglio di Stato, V, 5 agosto 2022, n. 6934* [Council of State, V, 5 August 2022, n. 6934]. Retrieved from www.giustizia-amministrativa.it/portale/pages/istituzionale/visualizza?nodeRef=&schema=cds&nrg=202110818&nomeFile=202208773_11.html&subDir=Provvedimenti

Iurascu, A. (2023). *How will the Adoption of Mandatory GPP Criteria Change the Game? Lessons From the Italian Experience, European Public Private Partnership Law Review 1*. Retrieved from https://sapiensnetwork.eu/new-article-mandatory-gpp-criteria-adoption/

Joint Research Centre, Institute for Prospective Technological Studies, Lee, P., Wolf, O., & Eatherley, D. (2016). *Revision of the EU Green Public Procurement Criteria for Food and Catering Services*, European Union Publications Office, Luxembourg. doi: 10.2791/099130.

Leal Filho, W., Caughman, L., Pimenta Dinis, M. A., Frankenberger, F., Azul, A. M., & Salvia, A. L. (2022). Towards symbiotic approaches between universities, sustainable development, and cities. *Scientific Reports*, *12*(1). doi: 10.1038/s41598-022-15717-2

Mohammed, A. M. S., Ukai, T., & Hall, M. (2022). Towards a sustainable campus-city relationship: A systematic review of the literature. *Regional Sustainability*, *3*(1). doi: 10.1016/j.regsus.2022.03.004

Olen, S. M., (2023). Civic sentinels. *Nature Sustainability*, *6*, 352–353. doi:10.1038/s41893-023-01103-x

Rodin, J. (2007). *The University and Urban Revival. Out of the Ivory Tower and Into the Streets*, University of Pennsylvania Press, Philadelphia.

Sirsi, E. (2019). Strumenti giuridici per l'alimentazione sostenibile nell'università [Legal instruments for sustainable nutrition in universities]. In Egidio, D., Fassio, F., Sirsi, E., & Tecco N. editor(s), *Le università e il cibo. Buone pratiche verso un'alimentazione sostenibile* [Universities and food. Good practices towards sustainable food], Celid, Torino.

Soini, K., Jurgilevich, A., Pietikäinen, J., & Korhonen-Kurki, K. (2018). Universities responding to the call for sustainability: A typology of sustainability centres, *Journal of Cleaner Production*, *170*, 1423–1432. doi: 10.1016/j.jclepro.2017.08.228

Swensson, L. F. J., Hunter, D., Schneider, S., & Tartanac, F. (2021). Public food procurement as a game changer for food system transformation. *The Lancet Planetary Health*, *5*(8), e495–e496. doi: 10.1016/S2542-5196(21)00176-5

Swensson, L. F. J., & Tartanac, F. (2020). Public food procurement for sustainable diets and food systems: The role of the regulatory framework. *Global Food Security*, *25*, 100366. doi: 10.1016/j.gfs.2020.100366

Università degli Studi di Torino. (2016). *Piano Strategico 2016–2020* [Strategic Plan 2016–2020], Università degli Studi di Torino. Retrieved from www.unito.it/sites/default/files/piano_strategico_2016_2020.pdf

Università degli Studi di Torino. (2021). *Piano Strategico 2021–2026* [Strategic Plan 2021–2026], Università degli Studi di Torino. Retrieved from https://pianostrategico2026.unito.it/

World Health Organization. (2004). *Global Strategy on Diet, Physical Activity and Health*. ISBN: 9241592222. Retrieved from www.who.int/publications/i/item/9241592222

References – European legislation

Eco-Design Regulation. (2019). *Commission Regulation (EU) 2019/2024 of 1 October 2019 Laying Down Eco-Design Requirements for Refrigerating Appliances With a Direct Sales Function Pursuant to Directive 2009/125/EC of the European Parliament and of the Council*. Retrieved from https://eur-lex.europa.eu/legal-content/EN/TXT/HTML/?uri=CELEX:32019R2024.

Energy Labelling Regulations. (2019). *Commission Delegated Regulation (EU) 2019/2018 of 11 March 2019 Supplementing Regulation (EU) 2017/1369 of the European Parliament and of the Council With Regard to Energy Labelling of Refrigerating Appliances With a Direct Sales Function*. Retrieved from https://eur-lex.europa.eu/legal-content/EN/TXT/HTML/?uri=CELEX:32019R2018

Single Use Plastic Directive. (2019). *Directive (EU) 2019/904 of the European Parliament and of the Council of 5 June 2019 on the Reduction of the Impact of Certain Plastic Products on the Environment*. Retrieved from https://eur-lex.europa.eu/legal-content/EN/TXT/HTML/?uri=CELEX:32019L0904&qid=1690703585345

9 Sustainable public food procurement in Italy

Minimum Environmental Criteria and definitions of sustainability

Andrea Appolloni, Wenjuan Cheng, Giorgia Masili, Daniele Binci and Mark Stein

Introduction

Public procurement has been widely recognised as a policy tool with potential to mitigate climate change, promote sustainable development and inspire innovation. In the food sector, sustainable public procurement is also increasingly recognised as a way of promoting a more sustainable food system and consumption of healthier food (Neto & Gama Caldas, 2018; Swensson et al., 2021). Unlike procurement within the private sector, public procurement is highly regulated. It is therefore very important that the regulatory framework should promote healthy and sustainable public food procurement.

Italy is a front-runner in the policy design and implementation of sustainable public procurement, with particular emphasis on environmental criteria, which is often referred to as Green Public Procurement (GPP) (European Parliament, 2017, pp. 79–80).

Italy has specific rules which create mandatory public procurement standards, the Minimum Environmental Criteria (MEC) for specific sectors and types of contracts [The Italian phrase is *Criteri ambientali minimi* – abbreviated as CAM]. This chapter presents a detailed discussion of the MEC which provides a comprehensive vision for sustainable public food procurement.

The National Action Plan on Green Public Procurement

In 2007, the Ministry of the Environment for the Protection of Land and Sea published the National Action Plan on Green Public Procurement (GPP NAP).[1] The Action Plan's purpose was to

> promote the dissemination of GPP to public bodies and facilitate the necessary conditions so that the GPP can achieve its full potential as an instrument for improving the environment.
>
> (Ministry of Environment & Protection of Natural Resources, 2007, p. 5)

The Action Plan defined national targets for GPP. It stressed that

DOI: 10.4324/9781003393023-11

public procurement should refer to the three pillars of sustainability (environmental, economic, and social), the social criteria including health and safety and being incorporated with the ethical criteria. In this regard, it would be more appropriate to use the term Sustainable Public Purchasing, or SPP, as well).

(Ministry of Environment & Protection of Natural Resources, 2007, p. 4)

Minimum Environmental Criteria (MEC)

The National Action Plan established Minimum Environmental Criteria to promote sustainable consumption in the public sector. The Italian government drew up these criteria in consultation with industry stakeholders and taking into account national market conditions. The use of these Minimum Environmental Criteria is required by Art. 18 of Law 221/2015 and by the 2016 Public Contracts Code (Legislative Decree 50/2016), which lays down the rules for all public procurement contracts (Botta, 2023). Article 34 of the 2016 code reiterates the importance of considering Minimum Environmental Criteria in preparing tender documents. All public sector tenders are expected to make use of the technical specifications and contractual clauses set down in the Minimum Environmental Criteria document. Green award criteria may be taken into consideration in tender documents which applied the criterion of the most economically advantageous bid. This approach has been confirmed under Article 57(2) of the 2023 Public Contracts Code (Botta, 2023).

Minimum Environmental Criteria are also expected to promote a thorough audit of contract compliance by providing for more effective verification methods and facilitating communication between the contractor and the public sector customer (Ministry of Ecological Transition, 2020).

The National Recovery and Resilience Plan (PNRR) set up new rules, which included the adoption of the "Do No Significant Harm" (DNSH) principle. Such a principle represented "the heart of a classification system of environmentally sustainable economic activities, which is applicable in several pieces of legislation in the area of sustainable finance" (Klika, 2022).

2020: Minimum Environmental Criteria for food & catering services

The Ministerial Decree of 10 March 2020 (Gazzetta Ufficiale No. 90, 4th April 2020) established Minimum Environmental Criteria for public procurement to identify the "best environmental design solution, product or service along the life cycle, taking into account market availability".

Article 2 of the Ministerial Decree defined Minimum Environmental Criteria for specific sectors of mass catering and food procurement. These criteria replaced the previous Criteria adopted in 2011 (Official Gazzette, 2011).

Mandatory Environmental Criteria (CAM) for collective catering and food

The CAM report (Ministry of Ecological Transition, 2020) is a detailed report (45 pages) which sets out the reasoning and evidence base justifying the Mandatory Environmental Criteria for collective catering services and the supply of foodstuffs. Page references in this section will refer to the CAM report.

Sectors

The new Criteria categorised mass catering service into the following sectors: a) schools and nurseries catering, b) offices, universities, and barracks and c) hospitals, welfare, social health and prisons (Art. 1, Gazzetta Ufficiale No. 90, 4th April 2020). Different criteria were envisaged for the different sectors, particularly in respect of minimum levels of organic food. The great distinction is between catering services oriented to children and young people and those intended for adults. Catering services for children and young people mainly related to nurseries, kindergartens, primary and secondary school. A smaller number of users came under other categories: hospital departments for babies and children and welfare, socio-medical and detention facilities. Catering services oriented to adults comprised offices, universities, barracks, prisons and hospitals – patients and staff (CAM report, pp. 36–40).

Market context

The report estimated that the collective catering market in Italy was worth around 6.5 billion euros annually for 1.5 billion meals and 110,000 employees. Approximately 62% was outsourced and the rest managed in-house (CAM report, p. 24). In 2015, 66% of school food was conventional and 97% of food in healthcare or employee canteens.[2]

The report also estimated that that 50% of the national catering market was covered by six large companies, each of which served between 28 and 36 million meals a year. The centralisation of procurement favoured a meal production model increasingly oriented towards large, centralised cooking centres replacing in-house public sector kitchens. These cooking centres tended to use very energy-intensive meal preparation techniques such as cook-chill or cook-freeze and to seek large-scale suppliers who could provide large quantities of highly standardised ingredients. Their system of supply tended not to be directed towards small and very small local producers, which are typical Italian organic producers. They are more likely to resort to foreign suppliers and/or buy from a few large national producers who can form a cartel and impose high selling prices (CAM report, p. 24).

Reducing meat and other animal protein

The CAM report emphasises that worldwide meat consumption is excessive – taking up most of the earth's fertile land for production of animal feed (p. 10).

Animal husbandry causes an unsustainable waste of resources, an impact on soil and water resources as well as strong repercussions on animal health, on the environment and on human health (CAM report, p. 9). The report underlines that purchasing organisations should ensure that menus have a lower consumption of animal proteins, while still complying with minimum nutritional requirements (CAM report, p. 27). It alluded to the importance of the Mediterranean diet – with relatively low meat consumption. Reference was made to the national guidelines for school catering which recommended meat one to two times a week and fish one to two times a week. This gives caterers latitude to offer plant-based food for up to three days a week (Health Ministry, 2010).

Organic food: minimum levels

The Criteria set out the requirement for minimum levels of organic or other quality marked foods. On the latter see further later on.

In schools and nurseries – and other services for children and young people – the requirement for organic or quality marked fruit and vegetables should be a minimum 50% by weight. In offices, universities and barracks – and other catering services for adults – they should be a minimum of 20% (p. 35).

Similar requirements applied to other products such as cereals, legumes, fruit juice and dairy products. Beef for children and young people was required to be 50% by weight organic or quality marked, compared to 20% for adults. The requirements for pork meat by contrast was only 10% for children and 5% for adults – reflecting much smaller availability of organic/quality marked pork meat. That for poultry meat for children was 20% by weight – also reflecting low availability (p. 36).

Environmental benefits of organic food

The CAM report emphasises that organic agriculture has a lower environmental impact per surface unit and promotes a better level of biodiversity and soil fertility (CAM report, p. 27).

It highlights the high level of surface and ground water contamination caused by herbicides and pesticides used within conventional agriculture and warns that the pesticide and herbicide levels judged acceptable by legislation could still be doing serious harm to human health and the environment (CAM report, p. 11).

The CAM report further asserts that scientific studies have shown that the ability of synthetic fertilisers to increase crop yields is declining while it has been found that careful management of appropriate agricultural practices (agroecology; polyculture; crop rotation; use of traditional cultivars) could further reduce the productive disadvantages of crops that do not use fertilisers and pesticides compared to those of conventional type. In the long run the organic method will guarantee food safety and nutritional quality while respecting human and environmental health (CAM report, p. 27).

Organic: Health benefits for children

The CAM report asserts that

> Numerous recent studies have in fact demonstrated how diets with a high consumption of organic foods are generally characterized by nutritional and environmental benefits.
>
> (CAM report, pp. 15–16)

Organic food is particularly beneficial for the health of infants and young children, because it has the lowest levels of pesticide residues (CAM report, pp. 16–17). A minimum of 50% organic food was envisaged for school catering and healthcare catering for infants and children (CAM report, p. 28). For adult users of collective catering the recommendation was for a minimum of 20% organic food.[3]

Organic: Increase in production and consumption

Italy had seen a very large increase in organic agricultural production in recent years. Between 2010 and 2017 the percentage of the cultivated agricultural area which was cultivated organically increased by 71%, reaching 15.4% (CAM report, p. 25). Between 2009 and 2018, the Italian organic market increased from 1.6 to 4.1 billion euros. This included a large increase in organic food in school kitchens (CAM report, p. 25).

Limited supply of Italian organic food

The CAM report acknowledges that, national production of organic food for some products and in some parts of Italy was not yet sufficient to cover the needs of public collective catering (CAM report, p. 27). From an estimate of the needs of school catering alone, compared with the data on national organic production, it could be seen that for some commodities national production was sufficient (for example, for eggs, some vegetables and fruit), while for other products there were difficulties with availability (such as poultry meat) (CAM report, p. 25).

The decision was therefore taken to lower the overall minimum percentage of organic food compared to the previous Minimum Environmental Criteria issued in 2010 and to give priority for organic food to catering sectors serving children and young people – who would have the greatest benefit from organic food.

Additional cost of organic food

The CAM report (p. 22) estimates that the cost of an average Italian school meal was 4.6 euros, and the average cost of food was 1.6 euros per meal. Postulating a 50% increase in the prices of organic ingredients compared to conventional food the increase in the meal price from replacing conventional with organic food, would be 0.8 euros. The report further suggests a number of ways in which this increase could be partially offset:

- Reducing energy consumption
- Implementing measures to reduce waste
- Not using fresh-cut and ready-to-eat products, which are much more expensive than fresh products
- Using cheaper recipes – such as pasta or rice with fish or meat sauce, legumes and sautéed vegetables and cheese

If the political will existed, the extra cost of organic food could be met with public funds (CAM report, p. 22).

Other quality marked foods

The CAM report (p. 28) stresses the importance of other quality marks. These include "Protected Designation of Origin" (PDO) and "Protected Geographical Indication" (PGI) products, which are recognised under EU regulation 510/2006 as having a particular product specification, originating in a specific place, whose characteristics are essentially due to a particular geographic environment. The report particularly stresses the importance of these quality marks in respect of beef, pig, or chicken meat (pp. 35–36). Under EU public procurement regulations, it is possible to specify that products purchased must comply with a PGI or PDO which can have the effect of directing business to Italian businesses which can supply these products (Schoenmaekers, 2021; European Committee of the Regions, 2018).

Short supply chains and Km Zero

The Criteria

> favor a more sustainable service, through the rewarding criterion of the short supply chain and the zero km, through which they also aim to support, as far as technically possible, local, and small organic producers. (The criterion, in order not to be detrimental to the principle of protection of competition, refers only to organic products).
>
> (CAM report, p. 30)

This provision is aimed at providing fairer compensation for farmers, reducing transport GHG emissions and enhancing organic food production. This is further likely to encourage conversion of agri-food production to organic and to support bio districts and other forms of association between producers, as envisaged by the national strategic plan for the development of the organic system.[4]

Long supply chains cause price increases without benefitting producers. With short, local, small-scale supply chains food lasts longer for the consumer since there are fewer steps and less chance of deterioration and consequent waste (CAM report, pp. 21, 30). Some studies have shown that short food chains may reduce food waste dramatically – bringing down the levels of rejects and waste in the stages prior to final consumption down to 5%, when normally these levels fluctuate between 30 and 50% (CAM report, p. 31).

Seasonal menus

Collective catering was encouraged to adopt seasonal menus in respect of fruit and vegetables and fish products. Fresh seasonal products are likely to be nutritionally superior. There are also environmental reasons for not buying out of season products. They are likely to be from an intensive production system or from heated greenhouses, with greater energy and water consumption or imported products, perhaps from countries where there are fewer controls on quality and effects on the environment of the production system, also generating environmental impacts due to transport from greater distances (CAM report, p. 30).

Stages of processing

The CAM report refers to five stages of processing:

- First range: traditional fresh fruit and vegetables
- Second Range: preserved fruit and vegetables in tins or jars
- Third Range: frozen fruit and vegetables
- Fourth Range: fresh fruit and vegetables washed, cut up packaged and ready for consumption
- Fifth Range: cooked fruit and vegetables, packaged and ready for consumption (CAM report, p. 12)

Pre-cut fruit and vegetables discouraged

From a nutritional point of view, pre-cut fruit vegetables (fourth range) are likely developing oxidative browning and lose original consistency after cutting. They are also likely to lose vitamin B and C content. There is a risk of bacteriological contamination between cutting and consumption (CAM report, pp. 13, 19) Contracting authorities were encouraged to forbid use of pre-cut fruit or vegetables, although they could make exceptions for specific circumstances – for example where there is no suitable kitchen space to cut fruit and vegetables on site (CAM report, p. 29).

Frozen products discouraged, with certain exceptions

The CAM report (p. 13) stresses the high environmental impact of frozen products associated with the "cold chain" – energy consumption by refrigerated storage and vehicles. It concedes however that for technical-operational reasons it can be desirable to purchase frozen fish and certain vegetables such as green beans or peas and frozen can provide an alternative in the absence of fresh product (CAM report, p. 29).

Pre-cooked products should be avoided

Collective kitchens were strongly advised to avoid the so-called fifth range products – dishes cooked and subsequently stored at a temperature of 0–3°C or frozen or vacuum-packed (CAM report, pp. 12–13). The CAM report stresses that these products have great environmental impacts (p. 13). Their production involves standardising shapes and sizes which creates additional food waste (CAM report, p. 13). Cultivation systems are likely to make high use of pesticides and to have high sowing density and farm specialisation, thereby promoting the spread of plant diseases.

Such products require enormous quantities of single-use packaging (CAM report, p. 29).

The CAM report further argued that use of these products was legally questionable.

> Among the essential activities of catering services is the preparation, including cooking, of the dishes offered. Is it therefore correct from a contractual point of view to entrust essential phases of the service such as preparation and cooking to third parties outside the subcontracting chain?
>
> (CAM report, p. 29)

The CAM report further argues (p. 29) that the ever more extreme artificialisation and standardisation of the foods and tastes that these products bring with them also have a negative effect on food education, especially for children, which is an important objective of collective public catering (p. 29).

Food waste

The CAM report refers to a high level of food waste in collective catering – 35–40% of meals produced in school catering and about 30% in hospital catering (CAM report, p. 13). The CAM require caterers to monitor food waste distinguishing between the different dishes and between unserved food and plate waste. Caterers should be required to carry out questionnaire surveys, to identify the causes of food waste and take actions to prevent waste. Uneaten food from collective catering should whenever possible be donated for charitable purposes – such as free distribution of food products. Uneaten food may be composted. It may also be used for animal feed (particularly in kennels and catteries rather than animals being raised for human consumption). Logistical costs should be minimised (CAM report, p. 32).

Packaging

The Criteria place strong emphasis on reducing packaging waste generated by public kitchens (CAM report, p. 32). Mass catering should provide water and drinks on tap or with returnable containers. All packaging should be reduced to a minimum by choosing refillable products or products that use "returnable" packaging. Packaging should ideally be recyclable, reusable or biodegradable and compostable.

Reusable tableware required

The Criteria require caterers to have reusable tableware because this is the environmentally preferable solution, even taking into account the consumption of water and energy for washing dishes. Disposable plastic tableware is extremely difficult to recycle for any purpose and usually ends up being incinerated (CAM report, p. 31).

Fish

The Criteria attempt to safeguard biodiversity by discouraging usage of over-exploited fish species which were listed as endangered. Overuse of mackerel and skipjack and yellowfin tuna in collective catering was a particular concern (CAM report, p. 11). It was further recommended that fishing areas should be limited to the Mediterranean and the North East Atlantic. For social reasons the FAO Eastern Central Atlantic fishing area was excluded – running along the coast of West Africa, where over-fishing has been a problem. An important ethical-social objective of the Criteria was to avoid subtracting previous food resources from poor populations with food security problems. European fleets fishing tuna have not only greatly reduced global stocks of this fish but subtracted a previous resource for local populations in countries like Senegal due to fishing methods used which involve the capture of many other species of interest to small local fishermen. Such local populations have been heavily dependent on protein from coastal fisheries, whose depletion have helped drive increasing migration from these countries (CAM report, pp. 22–23, 28).

The Criteria encouraged the use of less familiar species in school kitchens and suggested adopting a greater variety of fish recipes with a view to creating local fish supply chains This could include greater use of farmed fish – although there was a need to take account of environmental concerns relating to the use of pesticides and fishmeal in intensive fish farming (CAM report, p. 12).

Avoiding labour exploitation

The Criteria placed great emphasis on measures aiming to protect farm workers from exploitation.

> The agricultural sector is, as known, often subject to illegal work and "*caporalato*", a widespread illegal phenomenon of recruitment and brutal exploitation of laborers through intermediaries, the so-called "corporals" who hire manpower without respecting the labor regulations, withholding part

> of the salary and often without paying social security contributions, issuing untruthful pay slips, with unregulated working hours.
>
> (CAM report, p. 20)

Public procurers needed

> To combat this phenomenon and ensure that food produced through such despicable forms of exploitation, sometimes such as to cause loss of human life, is not served in public canteens.
>
> (CAM report, p. 20)

It was envisaged that companies bidding for food contracts would be required to set up a "structured dialogue" along the supply chains. This would examine the supply chains up to the farms of origin of the products to verify on site the methods of managing the work including labour-intensive operations such as harvesting. Bidders would be required to show that they had in place an agreement with a company with specialised knowledge of how to examine these matters through interviews directly with the workforce.

For the same purpose of avoiding the exploitation of the workers it was envisaged that companies bidding for food contracts would be given extra points if they purchased products from companies involved in the "Quality Agricultural Work Network" [Rete di lavoro agricolo di qualità]. This network was established by law in 2014 and created a register of companies which complied with all labour, social, public safety, tax, and other applicable laws (CAM report, p. 21).

Fair trade – tropical products

The Criteria (CAM report, p. 23) require that tropical products (such as coffee, sugar, and chocolate) should come from Fairtrade with specific certification such as that from the Fairtrade Labelling Organisation (FLO-Cert) or the World Fair Trade Organization. Whenever possible they should also be organic.

Vegetable Oils

With regard to vegetable oils other than olive and sunflower oil, purchasers are expected to avoid products which are produced from land with high biodiversity and carbon stocks (p. 30). This particularly refers to the need to avoid purchases of palm oil which has been produced through destruction of tropical rain forest.

Compliance checks

The Criteria envisaged a more effective approach to compliance checks to ascertain whether a contractor was actually implementing all the terms of their contract – for example relating to the percentages of organic food actually being used in meals. This could include on-site sample-based checks of purchase invoices and raw material transport documents (CAM report, p. 33).

Energy consumption

The Criteria stress the need to reduce energy consumption and related GHG emissions. New cooking equipment needs to meet energy efficiency requirements. Refrigerators and freezers need to be chosen using refrigerant gases with lowest global warming potential (CAM report, p. 32).

CAMs widely adopted within Italy

The Green Procurement Observatory carried out a questionnaire survey of 111 Italian municipalities – all the provincial capitals and received 88 responses. Its report concluded that for 2021 the adoption rate of Minimum Environmental Criteria within mass catering in 88 Italian municipalities was equal to 53.7% (Osservatorio Appalti Verdi, 2022). This registered a substantial increase compared to 2020, when the percentage was 45.6% (Osservatorio Appalti Verdi, 2021). The questionnaire survey, also conducted for 2022, assessed that the percentage of municipalities that adopted CAM for mass catering saw an increase to 75% (Osservatorio Appalti Verdi, 2023). It should be noted that the criteria used for food procurement in certain municipalities also included social issues.

Italian academic literature

A brief survey of academic literature about Italian public food procurement in Italy showed that certain Italian local authorities were pursuing sustainable food procurement policies which were consistent with the Criteria.

Cerruti et al. (2016) reported on work carried out in the City of Turin aimed at making the school catering service more sustainable. These included purchasing an increased proportion of organic food. To reduce "food miles", the City of Turin decided to restrict the place of origin for all major fruit and vegetables to the Piedmont region for the school year 2013/14. Other policies included acquisition of energy-efficient ovens and dishwashers, reduction in packaging; shifting from plastic to reusable dishes and redistribution of leftover food to social projects in the city". The project involved detailed calculations of GHG emissions from public kitchens for different food products.

Falasconi et al. (2015) investigated the reasons for food waste in six school canteens in Verona. The research concluded that this resulted from lack of attention to dietary habits, rigid food procurement specifications and deficiencies in menu composition, and meal presentation. Food waste could be reduced by adjusting the quantity of food when it was excessive; revising menus and information campaigns to promote better dietary habits. Where the: the school canteen donated food to an association helping vulnerable people, this should be located in the same area/neighbourhood to reduce the impact of transport and minimise food safety issues.

Galli et al. (2014) examined efforts to develop more sustainable school meal services in Pisa. The paper explores how this was shaped by the relationships between professionals and users of school meal services. These efforts were inspired by the

Minimum Environmental Criteria for public kitchens. Consideration was given to purchase of organic and seasonal products and the freshness of food, short supply chains, lower emissions, safe working conditions and minimisation of waste and packaging.

Lagorio et al. (2018) described how the Brusaporto municipality (Lombardy Region) implemented a strategy aiming at reducing food waste – the "Buon Samaritano" project. Students and families were educated to adopt better environmental behaviour and food culture. Unconsumed food was delivered to needy families.

Mazzà (2021) described how regional and local authorities within the Sardinia region worked together to develop sustainable food procurement policies in accordance with the National Action Plan for Green Public Procurement and the Minimum Environmental Criteria for public kitchens. This included implementing many of the measures advocated in the CAM report, such as organic food, seasonal fruit and vegetables, promotion of typical and traditional foods sustainable fishing; waste reduction through the avoidance of single-use cutlery and the use of recycled/recyclable packaging.

Tregear et al. (2022) analyses comparative case studies of sustainability in ten primary school catering services in five European countries. The Italian city of Parma had a catering contract in accordance with regional legislation promoting local, organic, typical and traditional products in school canteens. A minimum of 70% of the food served came from these sources. In the Italian city of Lucca, the regional guidelines and procurement contract specified the majority of foods for meal preparation should be organic.

Discussion and conclusions

The CAM report sets out the elements of a wide ranging sustainable public food procurement strategy – covering both environmental and social issues which in Italy has the force of being the policy of national government.

There are important similarities to organic food procurement strategies in Denmark which reflect many of the same themes. These are summed up in Mikkelsen & Sylvest (2012) and Sørensen et al. (2016).

- A long-term objective of replacing conventional with organic food in collective catering
- A target percentage of organic food
- A strong environmental case for promoting organic food – due to environmental pollution caused by pesticides, herbicides and fertilisers used in conventional agriculture
- Different techniques which have reduced costs and thereby generated savings to help finance the extra price of organic food
- Reductions in food waste
- Reductions in meat usage
- Seasonal menus
- Cooking from scratch whenever possible and phasing out of ready meals.

There are also numerous points of resemblance to the comprehensive sustainable food procurement strategy promoted by the UK voluntary sector initiative Food for Life Served Here – such as encouraging organic and local food, seasonal menus, cooking from fresh ingredients, reducing waste and meat, Fairtrade products and avoiding fish and palm oil from non-sustainable sources (Soil Association, 2019).[5]

As regards the special recognition given in the MECs to quality marks such as PDOs and PGIs, this can also be seen in other European countries where important food categories have been given this form of recognition, which enables public procurers to specify them in tenders and thereby open up public procurement opportunities for high-quality local foods (European Committee of the Regions, 2018). These quality marks are particularly concentrated in southern Europe (Dias & Mendes, 2018).

Other countries have seen public procurement initiatives relating to sustainable fish. This can be seen in Denmark and Sweden (Havibalance, 2023; Mattank, 2023).

There are public procurement initiatives elsewhere in Europe aimed at reducing palm oil usage – or limiting it to "sustainable palm oil". On Swedish policy see Upphandlingsmyndigheten (2023).

Throughout Europe there are public procurement initiatives aimed at encouraging purchasing of Fairtrade products (Fisher & Corbalán, 2013). This can include the designation of local authorities as Fairtrade cities which encourage purchasing by both citizens and public procurement. On the Swedish Fair Trade Cities, see Fairtrade Sverige (2021).

The Italian MEC requirement for avoiding labour exploitation stands out as having few parallels in other EU countries. The closest initiative is the Swedish KRAV standard – but this is a voluntary accreditation rather than a government requirement (KRAV, 2023).

Future research

There is abundant scope for future research in how these Minimum Environmental Criteria have affected public food procurement for mass catering in Italy.

Notes

1 The Ministry of the Environment for the Protection of Land and Sea was also called the Ministry of the Environment and the Protection of Natural Resources and was subsequently renamed the Ministry of the Environment and Energy Security. The National Action Plan on Green Public Procurement (GPP NAP) was drawn up in collaboration with the Ministries of Economy and Finance and Economic Development. The National Action Plan was based on law n.296/2006 and approved through the Interministerial Decree of 11th April 2008 (Gazzetta Ufficiale No. 107 of 8th May 2008).

2 The balance was organic food or another controlled or certified food.

3 The categories of mass catering designated to have a minimum of 20% of organic food were offices, universities, barracks and hospitals (adult staff and patients).

4 "A bio-district is a geographical area where farmers, citizens, tourist operators, associations and public authorities enter into an agreement for the sustainable management of

local resources, based on organic principles and practices, aiming at the fulfilment of the economic and sociocultural potential of the territory (Basile, 2017).

5 See other Chapters 1, 3 and 10, which discuss Food for Life.

References

Basile, S. (2017). *The Experience of Bio-Districts in Italy*. Retrieved from www.fao.org/agroecology/database/detail/en/c/1027958/

Botta, G. (2023). Italy: Leading the way towards mandatory sustainable public procurement through minimum environmental criteria. In Janssen, W., & Caranta R. (eds.), *Mandatory Sustainability Requirements in EU Public Procurement Law*, Oxford, Hart.

Cerruti, A. K., Contu, S., Ardente, F., Donno, D., & Becaro, G. L. (2016). Carbon footprint in green public procurement: Policy evaluation from a case study in the food sector. *Food Policy*, *58*, 82–93. doi: 10.1016/j.foodpol.2015.12.001

European Committee of the Regions. (2018). *Sustainable Public Procurement of Food*. Retrieved from https://cor.europa.eu/en/engage/studies/Documents/sustainable-public-procurement-food.pdf

Dias, C., & Mendes, L. (2018). Protected designation of origin (PDO) protected geographical indication (PGI) and traditional specialty guaranteed (TSG): A bibliometric analysis. *Food Research International*, *103*, 492–508. doi: 10.1016/j.foodres.2017.09.059

European Parliament. (2017). *Green Public Procurement and the EU Action Plan for the Circular Economy*. Retrieved 29 August, 2023, from https://data.europa.eu/doi/10.2861/3120

Fairtrade Sverige. (2021). *Bli diplomerad Fairtrade City och Region* [Become a Fairtrade City and Region]. Retrieved from https://fairtrade.se/kommun-och-region/bli-diplomerad-fairtrade-city-och-region/kontaktuppgifter-fairtrade-city-och-region/

Falasconi, L., Vittuari, M., Politano, A., & Segrè, A. (2015). Food waste in school catering: An Italian case study. *Sustainability*, *7*(11), 14745–14760. doi: 10.3390/su71114745

Fisher, E., & Corbalán, S. (2013). Fair trade and European public procurement: Legal principles and governance dynamics. *Social Enterprise Journal*, *9*(1), 11–27. doi: 10.1108/17508611311329985

Health Ministry. (2010). *Linee di indirizzo nazionale per la ristorazione scolastica* [National Guidelines for School Catering] (Official Journal no. 134 of 11 June 2010). Retrieved from www.salute.gov.it/imgs/C_17_pubblicazioni_1248_allegato.pdf

Galli, F., Brunori, G., Iacovo, F. D., & Innocenti, S. (2014). Co-producing sustainability: Involving parents and civil society in the governance of school meal services. A case Study from Pisa, Italy. *Sustainability*, *6*(4), 1643–1666. doi: 10.3390/su6041643

Havibalance. (2023). *Baggrund* [Background]. Retrieved from www.havibalance.com/baggrund

Klika, C. (2022). *Sustainable Finance in the EU: Implementing the Commitment to 'Do No Significant Harm'*. European Institute of Public Administration. Retrieved from www.eipa.eu/publications/briefing/sustainable-finance-in-the-eu-implementing-the-commitment-to-do-no-significant-harm/

KRAV. (2023, 8 June). *This Is KRAV: Better Working Conditions*. Retrieved from https://www.krav.se/en/this-is-krav/better-working-conditions/

Lagorio, A., Pinto, R., & Golini, R. (2018). Food waste reduction in school canteens: Evidence from an Italian case. *Journal of cleaner production*, 199, 77–84. doi: 10.1016/j.jclepro.2018.07.077

Mattank. (2023). *Hållbar sjömat – nya möjligheter för offentliga måltider* [Sustainable Seafood – New Possibilities for Public Meals]. Retrieved from https://mattanken.se/

mattanken/merkunskap/samladkunskapochmaterial/material/hallbarsjomatnyamojlighete rforoffentligamaltider.5.63a7ae421865f604b9bac183.html

Mazzà, L. (2021). Planning and networking for green procurement in the food Sector: The case of the region of Sardinia, Italy. In Swensson, L. et al. (eds.), *Public Food Procurement for Sustainable Food Systems and Healthy Diets*, Rome, FAO & Bioversity International, Chapter 27, Vol. 2, 259–267. doi: 10.4060/cb7969en

Mikkelsen, B. E., & Sylvest, J. (2012). Organic foods on the public plate: Technical challenge or organizational change? *Journal of Foodservice Business Research, 15*(1), 64–83. doi: 0.1080/15378020.2011.650541

Ministry of Ecological Transition. (2020). *Relazione di accompagnamento CAM Servizio di ristorazione collettiva e fornitura derrate alimentari* [Accompanying Report Mandatory Environmental Criteria Collective catering Service and Supply of Foodstuffs]. (DM 10 marzo 2020, G.U. 4 aprile 2020, n. 90). Retrieved from www.mase.gov.it/sites/default/files/archivio/allegati/GPP/2020/relazione_accompagnamento_cam_ristorazione_aprile2020.pdf

Ministry of Environment & Protection of Natural Resources. (2007). *Action Plan for the Environmental Sustainability of Consumption in the Public Administration Sector. Or National Action Plan on Green Public Procurement – GPP NAP (English Version)*. Retrieved from www.mase.gov.it/sites/default/files/archivio/allegati/GPP/all.to_21_PAN_GPP_definitivo_EN.pdf

Neto, B., & Gama Caldas, M. (2018). The use of green criteria in the public procurement of food products and catering services: A review of EU schemes. *Environment, Development and Sustainability, 20*(5), 1905–1933. doi: 10.1007/s10668–017–9992-y

Official Gazzette. (2011). *Gazzetta Ufficiale della Repubblica Italiana, 21–9–11 Piano d'azione per la sostenibilità ambientale dei consumi nel settore della Pubblica Amministrazione ovvero Piano d'Azione Nazionale sul Green Public Procurement (PANGPP) Criteri Ambientali Minimi per il servizio di ristorazione collettiva e la fornitura di derrate alimentary* [Action Plan for the Environmental Sustainability of Consumption in the Public Administration Sector or National Action Plan on Green Public Procurement (PANGPP) Minimum Environmental Criteria for the Collective Catering Service and the Supply of Food Products]. Retrieved from www.bosettiegatti.eu/info/norme/statali/2011_dm_25_07_cam_ristorazione.pdf

Official Gazzette No. 90. (2020). *Gazzetta Ufficiale della Repubblica Italiana, 4–4–2020 Criteri ambientali minimi per il servizio di gestione del verde pubblico e la fornitura di prodotti per la cura del verde* [Minimum Environmental Criteria for the Public Green Management Service and for the Supply of Green Care Products]. Retrieved from www.bosettiegatti.eu/info/norme/statali/2020_dm_10_03_cam_verde_pubblico.pdf

Osservatorio Appalti Verdi. (2021). *I Numeri del Green Public Procurement in Italia Rapporto 2021* [The Numbers of the Green Public Procurement in Italy Report 2021]. Retrieved from www.fondazioneecosistemi.org/wp-content/uploads/2021/10/Rapporto_I-numeri-del-Green-Public-Procurement-2021.pdf

Osservatorio Appalti Verdi. (2022). *I Numeri del Green Public Procurement in Italia Rapporto 2022* [The Numbers of the Green Public Procurement in Italy Report 2022]. Retrieved from www.forumcompraverde.it/wp-content/uploads/2022/10/I-numeri-del-Green-Public-Procurement_Rapporto-OAV-2022.pdf

Osservatorio Appalti Verdi. (2023, 23 May). *I primi risultati 2023 del monitoraggio civico dell'osservatorioappaltiverdi* [The First 2023 Results of the Civic Monitoring of the Osservatorioappaltiverdi]. Retrieved from https://www.appaltiverdi.net/i-primi-risultati-2023-del-monitoraggio-civico-dellosservatorioappaltiverdi/

Schoenmaekers, S. (2021). Public procurement, culture and Mozzarella: 'Que Dici?'. *European Procurement & Public Private Partnership Law Review*, *16*(3), 205–219. doi: 10.21552/epppl/2021/3/6

Soil Association. (2019). *Food for Life Served Here: Handbook 2019 – Schools*. Retrieved from www.soilassociation.org/media/20184/fflsh-standards-handbook-schools.pdf

Sørensen, N. N., Tetens, I., Lassen, A. D., & Løje, H. (2016). *Organic Food Conversion in Danish Public Kitchens: The Effects of the Danish Organic Action Plan 2020 on Organic Public Procurement and Wellbeing at Work*. Søborg: National Food Institute, Technical University of Denmark. Published PhD thesis. Retrieved from http://orbit.dtu.dk/files/124231485/PhD_Thesis_Nina_N_rgaard_S_rensen_print_final.pdf

Swensson, L. F., Hunter, D., Schneider, S., & Tartanac, F. (2021). Public food procurement as a game changer for food system transformation. *The Lancet Planetary Health*, *5*(8), e495–e496. doi: 10.1016/S2542-5196(21)00176-5

Tregear, A., Aničić, Z., Arfini, F., Biasini, B., Bituh, M., Bojović, R., . . . & Tsakiridou, E. (2022). Routes to sustainability in public food procurement: An investigation of different models in primary school catering. *Journal of Cleaner Production*, *338*, 130604. doi: 10.1016/j.jclepro.2022.130604

Upphandlingsmyndigheten. (2023, 28 June). *Hållbarhetskriterier för ökad efterfrågan på mer hållbar palmolja och fodersoja* [Sustainability Criteria for Increased Demand for More Sustainable Palm Oil and Forage Soy]. Retrieved from www.upphandlingsmyndigheten.se/nyheter/2023/nyhet-publicering-av-reviderade-kriterier-for-palmolja-och-fodersoja/?utm_campaign=unspecified&utm_content=unspecified&utm_medium=email&utm_source=apsis-miljomalsnytt

Electronic references

www.codiceappalti.it/Home/Legge/?legge=Italian_Procurement_Code
https://eng.sinu.it/larn/
www.gazzettaufficiale.it/
https://gpp.mite.gov.it/Home/Cam
https://gpp.mite.gov.it/Home/PianoAzioneNazionaleGPP#1

10 Designing and measuring sustainability and social impacts of food service

Maurizio Mariani, Robin Gourlay and Giulia Caddeo

Introduction

In pursuing sustainable development goals and the Green Deal, there is a growing need to comprehensively assess the environmental and social impacts of public food service. As a response, new tools and indicators have emerged that provide a more nuanced understanding of the effects on both the environment and society.

The food industry in the EU has a substantial carbon and water footprint. Moving towards a more sustainable and healthier food service can significantly reduce emissions and pressure on water resources.

While decision-makers are aware that food affects the health of citizens, they are less aware of its impact on the world's resources. Thus, there is a need for tools that raise awareness of the potential resource savings from sustainable and healthy diets.

As the world strives for sustainability and positive social impact, traditional indicators often fail to capture the complexity and interdependencies of environmental and social systems. The tools described in what follows – carbon footprint, water footprint, social return on investment (SROI) and Local Multiplier – provide a more comprehensive and nuanced understanding of the impacts of human activities.

In summary, environmental indicators like carbon and water footprints, along with tools such as SROI and the local multiplier (LM3), offer a comprehensive framework to assess the impact of actions on both the environment and society. These tools foster sustainable practices, informed decision-making and collaboration towards a more equitable and resilient future.

The SROI indicator originated at the end of the 1990s in the UK and was initially used in Anglo-Saxon countries has spread thanks to the creation of the "Social Value international" network (www.socialvalueint.org/). It is currently used in several European countries, particularly in Italy and France, although to date we have not found any case studies belonging to the catering sector in these countries.

This chapter delves into four indicators: carbon footprint, water footprint, social return on investment (SROI), and Local Multiplier (LM3).

DOI: 10.4324/9781003393023-12

Designing the service

Designing a sustainable and healthy meal composition process for a school canteen service involves careful planning, ingredient selection, and menu design to ensure that the meals provided are nutritious, environmentally friendly and appealing to students. Here is a non-exhaustive step-by-step breakdown of the process:

Menu planning and design

Start by creating a diverse menu that includes a variety of food groups such as fruits, vegetables, whole grains, lean proteins, and dairy (or dairy alternatives). Aim for balance and variety in terms of flavours, textures and cooking methods to keep the meals interesting for students.

Nutritional guidelines

Follow national and international nutritional guidelines based on the dietary needs of students in different age groups. These guidelines should consider calorie requirements, nutrient recommendations and dietary restrictions.

Local and seasonal ingredients

Prioritise the use of local, seasonal and organic ingredients. This reduces the carbon footprint associated with transportation and supports local farmers. Seasonal ingredients are often fresher and more flavourful, enhancing the overall quality of the meals.

Plant-based options

Include various plant-based meal options to promote sustainability and cater to students with vegetarian or vegan preferences. Legumes, nuts, seeds, tofu and whole grains can be excellent plant-based protein sources.

Reducing food waste

Plan portion sizes carefully to minimise food waste. Implement strategies like offering smaller portion options or allowing students to choose their portion sizes. Utilise leftover ingredients creatively in other dishes to minimise waste.

Mindful cooking methods

Opt for cooking methods that require less energy, such as steaming, baking and grilling, instead of frying. Avoid overcooking to preserve the nutritional value of the food.

Diverse beverage options

Provide water as the primary beverage option and encourage students to stay hydrated. Offer water stations or reusable water bottles. Limit sugary drinks and provide healthier alternatives like unsweetened fruit juices, herbal teas or infused water.

Allergen awareness

Clearly label allergens on menus and educate staff about preventing cross-contamination for students with food allergies.

Education and engagement

Educate students about healthy eating and sustainability through workshops, presentations and informative displays. Involve students in menu planning to gather feedback and incorporate their preferences.

Continuous improvement

Regularly review and assess the menu to identify areas for improvement. Keep up with food trends and nutritional research to enhance the offerings. Adapt the menu based on seasonal availability and student preferences.

Collaboration and partnerships

Collaborate with local farmers, suppliers and nutritionists to ensure the best quality ingredients and up-to-date nutritional information.

Feedback loop

Establish a system for collecting feedback from students, parents and staff regarding meal quality, variety and sustainability efforts. Use this feedback to refine the menu and overall approach.

By following these steps, a school canteen can create a sustainable and healthy meal composition process that promotes the well-being of students while also contributing positively to the environment to fight climate change and boost the local economy.

Taking into consideration these steps as a building block will help decision maker to get a large picture and creating the base to measuring the real impact of their decision through a systemic vision of this service.

Measuring environmental sustainability

Measuring the catering service's sustainability is not such a simple task. We can begin with simple environmental indicators such as the energy (gas/electricity) consumed for the production of a meal, measure the litres of water needed, as well as the production of waste per product category, in the three different phases of the service, Production, Distribution, Consumption.

We can estimate CO_2 emissions related to primary logistics and measure fuel consumption in secondary logistics (last mile), thus obtaining a dashboard of indicators useful to define a meal's sustainability.

Nevertheless, for all these indicators measured downstream of the service, we will discover that they barely account for 20% of the total impacts, the main impact remains related to the ingredients used, to the source of protein used, and their geographic origin i.e. upstream of the service itself.

Thanks in part to several projects funded by the European Commission, there are now multiple tools available to calculate the environmental impact of the food we consume.[1] The simplest and most effective of these tools is indicating the carbon footprint (CF) and the water footprint (WF).

Carbon footprint is a term that describes the amount of greenhouse gases (GHGs) emitted by all the processes involved in producing a good or service (GHG Protocol, 2022). For goods, the carbon footprint of a product is the amount of CO_2 equivalents produced per unit of production. This unit could be a weight measurement, like a kilo or a production unit such as a wine bottle or a pot of yogurt. When calculating a product's carbon footprint, all stages from raw materials to waste disposal are considered.

According to ISO 14040:2006, the process of analysing the entire life cycle of a product is known as "Life Cycle Analysis". It involves calculating the total greenhouse gas emissions released during its creation, transportation, material production (including fertilisers and packaging), livestock or soil, and waste management.

The water footprint (Hoekstra & Chapagain, 2004) is an indicator of water use in relation to consumer goods and is similar to the concept of ecological and carbon footprints. The water footprint of a product is the volume of fresh water used to produce the product along the production chain.

As the water footprint is influenced by local climate, hydrology, crop types and other environmental pressures in the same water basin, it is a geographically explicit indicator. Therefore, the water footprint can significantly vary for the same food item depending on its place of production.

During the design phase of a dish or menu, utilising carbon and water footprint databases can provide a better understanding of its impact. It is no longer enough only to consider calories and nutrients; protein sources, whether animal or vegetable, must also be taken into account.

Let's delve into the details of setting measurable targets regarding their impact on climate change mitigation and efficient water use.

To calculate the daily CF limit, the figure expressed in the Guidelines for a Healthy Italian Diet (Rossi et al., 2022) was used in the study "Healthy Italian Dietary Pattern to Achieve 2030 Climate Target" (Ferrari et al., 2023) to estimate the CF equivalent to 2.3 KgCO2/capita/day on an average energy intake (males and females) of 2250 Kcal.

Obviously, this value can be related to a caloric requirement, determined for individual age/sex groups to be parameterised according to the age/sex of the food service users.

The value of 2.3 KgCO2/capita/day, for example, can define the limit to contain the environmental impact in terms of greenhouse gas emissions as much as possible.

For the calculation of the daily limit of WF, the value estimated on the reference diet published by Capone et al. (2013) equivalent to 2642 L/capita/day on an average energy intake (males and females) of 2250 Kcal, can be considered as a limit value to contain as much as possible the environmental impact in terms of freshwater consumption.

The distribution during the day of the percentage of calories and nutrients in the different moments, such as breakfast, lunch and dinner, can be done according to what is laid down for example in the Guidelines for a Healthy Italian Diet (CREA, 2018), thus distributing the CF and WF over the individual moments and thus establishing a measurable target to aim for.

The evidence emerging from various scientific research and experimentation, including those carried out in the European "Su-Eatable life" project, shows us the way forward to change the model and establish measurable goals to promote a healthier and more sustainable diet also in public food service.

All this may sometimes not be enough to push decision-makers to promote a sustainable diet in public procurement of catering services, which is why it is necessary to measure also the social impact through other tools and methodologies such as the Social Return on Investment (SROI) or by using the Local Multiplier (LM3).

Measuring social impacts

Social Return on Investment (SROI) and Local Multiplier (LM3) are frameworks for making judgements about how to account for value. They include a methodology and a set of principles. Establishing any return figure over 1 is obviously good and can validate a sustainable development agenda for change. In the example given using SROI in this chapter the return is between 1 and 6.

Is evaluation important?

Implementing a more sustainable public food system is an ambition most would agree with but in an era when local governments are under pressure to cut costs within reduced budgets, making the case for investment of any kind can be challenging if it cannot demonstrate immediate cost savings, early benefits or guarantee of a return in a predictable number of years.

A sustainable school food system can empower consumers by allowing access to food education, to good food and healthy habits, connecting with a sense of place, supporting social inclusion and ecological sustainability as well as delivering economic development through the creation of markets for local producers.

While public food service can deliver multiple dividends in social, economic and environmental terms, accounting for this broader concept of value over time requires a methodology which is able to articulate this.

School food is not generally viewed in terms of the broader value offered and is perceived as a cost burden by the local government. If there is no calculation or assessment of the short- and long-term benefits described earlier, then interventions designed to achieve a sustainable public food service are less likely to survive with political support.

In the public realm, it is vitally important to adhere to more than good intentions of what might be achieved by an intervention – such as creating a sustainable public food system for schools, hospitals and indeed anywhere that public money is involved. The activity needs to be in a position to clearly communicate the potential impact in more than financial terms.

For example, school food is in fact a health and education service with the power to create and sustain market opportunities for local businesses and food producers. As such, it requires a comprehensive assessment in order to understand its significant contribution to the life and well-being of a young person and in the wider community, contemporaneously and sustained into the future.

How to identify and quantify impact

Social Return on Investment (SROI) methodology accounts for the triple bottom line of social, economic and environmental value that results from the activity.

It is considered to be the foremost framework to measure value beyond simple financial characteristics and bring to the fore the key components of a healthier and more sustainable public food service system. The SROI methodology has the advantage of identifying the potential outcomes and calculates a monetary value attributed to the impact of activity as triple bottom-line assessment in social, economic and environmental terms.

The analysis can take many different forms. It can encompass the social value generated by an entire organisation or focus on just one specific aspect of its work. It can be carried out as an in-house exercise or led independently by an external researcher.

Types of SROI

1. Evaluative, which is conducted retrospectively and based on actual outcomes that have already taken place.
2. Forecast, which predicts how much social value will be created if the activities meet their intended outcomes.

The principles

The social return on investment (SROI) model provides an evaluation method for understanding, measuring and reporting on change, and the value created by an organisation or activity. It examines the social, economic and environmental impacts arising from the organisation's work, and attributes a value based on common accounting and investment appraisal methods, in order to estimate its financial value.

The SROI methodology was first developed in the USA and has been adapted for a European and UK context to take account of differences in accounting practices, and to increase the involvement of stakeholders in the analysis.

SROI is based on a set of principles:

- **Stakeholder perception**
    - Understand the way in which the organisation creates change. through dialogue with stakeholders
- **Scope and materiality**
    - Acknowledge and articulate all the values, objectives and stakeholders of the organisation before agreeing which aspects of the organisation are to be included in the scope; and determine what must be included in the account in order that stakeholders can make reasonable decisions
- **Understand change**
    - Articulate clearly how activities create change and evaluate this through the evidence gathered
- **Comparative**
    - Make comparisons of performance and impact using appropriate benchmarks, targets and external standards
- **Transparency**
    - Demonstrate the basis on which the findings may be considered accurate and honest; showing that they will be reported to and discussed with stakeholders
- **Verification**
    - Ensure appropriate independent verification of the account
- **Financial proxies**
    - Use financial proxies for indicators in order to include the values of those excluded from markets in the same terms as used in markets
- **The design and delivery**
    - An SROI analysis involves a series of set stages and standards
- **Boundaries**
    - Defining the scope of the work

- **Stakeholders**
  - Identifying and mapping objectives
- **Impact mapping**
  - Analysis of inputs, outputs and outcomes
- **Indicators**
  - Identifying the evidence base for impacts
- **Data**
  - Collecting required information
- **Model and calculate**
  - Financial modelling of social return
- **Testing**
  - Sensitivity analysis
- **Present**
  - Results
- **Verification**
  - Peer review

SROI can help make the organisation more sustainable by

- Highlighting the social, economic and environmental impacts of the service
- Raising the organisation's profile and reputation
- Underpinning the case for funding
- In the appropriate context, making catering tenders more robust and sustainable

Conclusion

It is becoming clear to policymakers, parents, educators, and practitioners that a school food service has the potential to deliver multiple dividends that can significantly advance the sustainable development agenda at national and local levels and enhance the reputation of the organisation. This requires a credible evaluation methodology such as SROI.

Case study

In August 2004 East Ayrshire Council in Scotland set off a quiet revolution in school meal procurement when it established a pilot project at one school to provide organic, local, and fresh food expanded to create one of the first sustainable school meals based on healthy, unprocessed, local and organic ingredients (Lancaster & Durie, 2008). From the pilot school in 2004, by the school year 2007/8 the initiative covered 26 of the authority's 43 schools and it still forms the basis on which the service is delivered in 2023, now across 44 locations serving around 5,500 meals a day, 3,600 of which are free school meals.

The primary objective of the pilot scheme was to "improve the health of primary school children and to educate children on food". Everything else was secondary in the project's early stages, but an appreciation of environmental sustainability, community regeneration, employment, and other economic and social impacts all played an important part in the decision to continue the pilot.

Robin Gourlay, the then head of facilities said, that they

> wanted, as a council, to see how sustainable we could make a school meal. We knew from our own research that local and organic food was the right approach; that there were things we could do as a council to reduce our carbon footprint; and that by using the findings from the social return on investment (SROI) study, by localising the food supply chain to using local suppliers within 40 miles, for every £1 we invested in food, the SROI was £6. Our vision was to achieve returns on health, energy use, food-miles, have tastier and fresher food and more. We wanted to provide appealing and imaginative school meals menus based on unprocessed, local and a proportion of higher animal welfare or organic ingredients. Using fresh food reduces packaging and a switch to organic reduces the environmental impact of school meals.

A significant issue was that there was limited scope for its existing large-scale suppliers to provide higher animal welfare, organic or fresh products at the volumes needed. A switch to developing relationships with local suppliers, and on configuring contracts on a scale which made it possible for smaller suppliers to bid for them was a key strategy. All of the procurements implemented complied with the full rigour of EU regulation.

The initiative has won many awards and international recognition, in addition to holding the Soil Association's Food for Life Gold Standard consistently for 15 years.

Previous evaluations recognised a range of social, environmental and health benefits. An evaluation commissioned by the Scottish Government reported that: "Further evidence of the monetary value of the economic, social and environmental impacts of local procurement will be crucial in encouraging local authorities to adopt new practices".

The SROI social return on investment study was also commissioned to add to such evidence to investigate the benefits created for stakeholders by the activity and to place financial values on outcomes achieved.

Result

SROI identified a range of environmental, economic, health and other outcomes. Making reasonable assumptions and applying recognised financial values to each of these outcomes demonstrated that the initiative created over £500,000 of value for stakeholders.

The additional investment by East Ayrshire Council in the period (above the cost of providing meals before the intervention) was 12.9p per school meal, or £70,838.

The social return on investment Index was 6. This meant that for every additional £1 invested in the initiative over £6 of value was created in economic, social, environmental and other outcomes.

The Sensitivity Analysis was carried out on varying major assumptions. This indicated the value of the return is unlikely to be below £3 for every £1 invested.

Stakeholders

These were school children, parents, local suppliers, teaching, catering and procurement staff, NHS and the Community Planning Partnership (who stood as a proxy for 'the environment') and identifying what changed for them.

Information collected came from, parent and child consultations, suppliers and staff surveys, interviews with others; detailed records on uptake of meals and results from other evaluation studies.

Value change

- Original research into valuing environmental outcomes e.g. a unit cost per food mile avoided
- Research modelling avoided healthcare costs within the National Health Service due to reduced childhood obesity
- Survey with suppliers looking at the impact on their businesses e.g. new land brought into organic production

Social outcomes

Health: £77,000 (15%)

School children

- Improved health
- Healthy school meals
- Better tasting school meals
- Greater understanding and appreciation of the local environment

There was also evidence that fewer children were eating junk food but this was not followed up.

Parents

- Improved health of children, Improved relationship with school
- Perception that Council was making good use of public money

- Healthy eating habits established
- Greater understanding and appreciation of the local environment

Economic outcomes

Economic: £258,000 (51%)

Local suppliers: More secure businesses; Greater access to other contract opportunities; Increased employment of staff; Increased organic production; Higher profile in the local community

Environmental

Environment: £98,000 (19%)

- Reduced food miles leading to less CO_2 – this was found to account for 95% of environmental value
- Reduced food miles leading to less environmental damage
- Reduced packaging waste going to landfill
- Improved externalities from organic production: 1% of value, suggesting more food could be organic

Local Multiplier LM3

LM3 is another methodology that enables an organisation to measure its economic impact by analysing any expenditure on any geographic area using current data.

The tool can be used by **the public sector** to provide objective evidence of community benefit, sustainability measurement and social value.

Local Multiplier 3 (LM3) was also developed by the New Economics Foundation as an understandable way of measuring local economic impact. It is designed to help think about local money flows and how their organisation can practically improve its local economic impact, as well as influence the public sector to consider the impact of its procurement decisions.

LM3 measures how income entering the local economy then circulates within it, across three "rounds" of spending.

An economic multiplier effect describes the impact that spending has in the economy, taking into consideration knock-on effects. The measuring process starts with a source of income and follows how it is spent and re-spent within a defined geographic area.

A higher proportion of money re-spent in the local economy means a higher multiplier effect because more income is generated for local people. More income retained locally, or nationally, means more jobs, higher pay and more tax revenue for government, all of which may lead to better living standards.

The model has been refined and improved based on work with all 26 local authorities in the Northeast of England, UK with over £3 billion of annual spending with 140,000 suppliers. This work formed an important step in influencing national procurement policy and areas of legislation such as Social Value Act and National procurement and sustainability policies.

Case Studies: Local economic multipliers of case school supply chains

Figure 10.1 illustrates the analysed cases of the Strength2Food Project. The project report *Strengthening European Food Chain Sustainability by Quality and Procurement Policy, D.6.3 Final Report* evaluates the environmental, economic and social impacts of different models of Public Sector Food Procurement (PSFP) in a school context (Strength2food, 2019).

In each of five countries of the research (Croatia, Greece, Italy, Serbia, UK), a pair of case studies was undertaken, each case study representing a contrasting model of PSFP. Each case model consisted of the supply chain providing meals to a sample of five schools in the case (four schools in the Serbian cases). Primary and secondary data were then collected to evaluate the sustainability impacts of these meals services. In four countries (Croatia, Greece, Serbia, UK), the paired cases comprised one 'LOW' model, where contract awards were made mostly or entirely on the basis of lowest price, and one 'LOC' model, where either the contract award criteria referred to local sourcing, or in practice the chain consisted of a proportion of local suppliers. In Italy, where by regional law a minimum of 70% of food procured for school meals must come from organic agriculture, integrated production or typical and traditional products, the two cases were LOC-ORG (a model

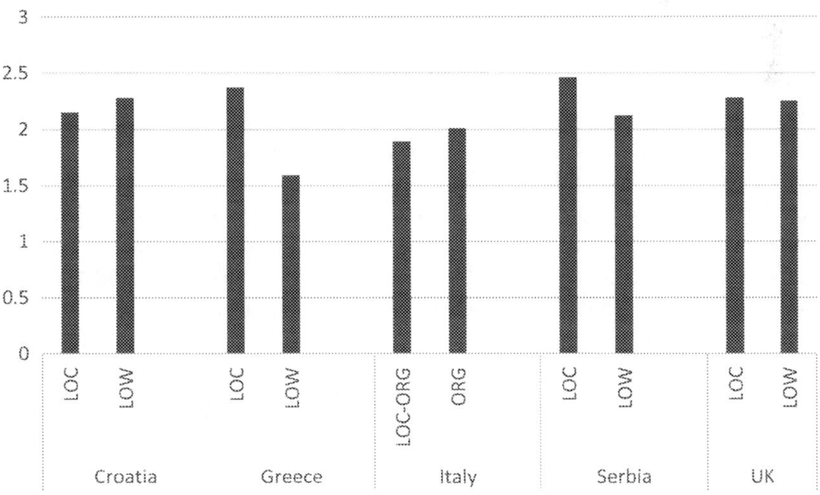

Figure 10.1 Local Economic Multiplier (LM3) ratios for Public Sector Food Procurement case models

Source: Adapted from Strength2Food, 2019, Figure 7, p. 34)

operating according to this law) and ORG (a model in which the contract primarily referred to organic sourcing).

To concretely describe the use of the LM3 criteria, we will provide data and details from the Serbian and Italian cases.

LM3 ratios for PSFP case models

The graph shows the local economic multiplier effect for different LOC cases. Serbia has the highest LM3 ratio of 2.46, meaning that for every 1.00 spent from the school meals budgets, an additional 1.46 was generated for the local economy within 15 km radius of each school.

Another interesting situation is the one of Italy, in which the dichotomy is not LOC/Low, but LOC-ORG/ORG cases. In Italy, in fact, whereby regional law a minimum of 70% of food procured for school meals must come from organic agriculture, integrated production or typical and traditional products, the two cases were LOC-ORG (a model operating according to this law) and ORG (a model in which the contract primarily referred to organic sourcing).

For three case pairs (Greece, Serbia and UK), the LOC PSFP models had higher LM3 ratios due to a greater proportion of supplier budgets being spent on local firms. However, a contrary result was found for the case pairs in Italy and Croatia, with the LOC case models having a smaller LM3 ratio than their counterparts. In Italy cases, the reason relates to the *de facto* degree of localisation in Parma LOC-ORG case.

Specifically, it was noted that although Parma municipality had the ambition to encourage local sourcing, the broad definition of local in the contract, and lack of minimum threshold, placed little obligation on the caterer to do so in practice. In fact, it was found that only 18% of the LOC-ORG supplier budget was spent on local firms, compared with 68% of ORG supplier budget. Hence the reason for the smaller LOC-ORG ratio. The result highlights an important point about the gap between ambition and reality of sustainability outcomes in PSFP, when contract criteria are not specified in the strongest way.

To evaluate economic impacts, data have been gathered on the flows of expenditures from meals service budgets, and staff/supplier locations, in order to estimate the local economic multiplier (LM3) effect of the meals services. Across the cases, the highest LM3 ratio calculated was 2.46 (Serbia LOC), and the lowest 1.59 (Greece LOW). The ratios indicated that in the highest case, every 1.00 spent from the school meals budgets generates an additional 1.46 for the local economy, whereas the additional value is only 0.59 in the lowest case. In terms of within-case pair differences, the results were as expected for three case pairs (Greece, UK and, precisely, Serbia), whereby ratios for LOC cases exceeded those of their counterparts, due to their proportionately higher expenditures on local suppliers. In Italy, as well Croatia, LOC models gave smaller LM3 ratios than their counterparts. For Italy, the explanation is that despite the municipal ambition to encourage local sourcing in the LOC-ORG case, there was a *de facto* low budget spend on local suppliers in this case. The result highlights how important it is for contracting authorities to translate sustainability goals into specific and measurable contract criteria, in order to truly influence procurement practices and economic multiplier effects.

Data also gathered from secondary sources, and from supplier interviews, to estimate the economic value of the contracts to suppliers. Overall, it was found that suppliers to the PSFP cases were a mix of large and small firms, indeed ranging in extremes from local microbusinesses (2 employees, turnover of €40,000) to very large national/international enterprises (2,000 employees turnovers of > €200million). However, in the vast majority of instances, the school meals contracts of the case models represented only very small, or negligible proportions of suppliers' total businesses, and these had not contributed directly to the winning of new business for those suppliers. The exceptions to this were two of the private catering firms (UK LOC caterer, Italy ORG caterer), and a handful of smaller firms in other cases. Nevertheless, in interviews, suppliers rated their involvement in the PSFP contracts positively, as a steady and complementary area of business. Also, the results possibly underestimate the value of PSFP contracts, as a whole, to the suppliers in the chains, as many were engaged in fulfilling multiple contracts. There were no notable differences found in the economic value indicators between the cases in each pair.

The report concludes with a range of recommendations to key authorities and decision-makers on how to enhance the sustainability outcomes of PSFP models. Although adoption of a localised procurement model can promote positive local economic multiplier effects and greater social connectedness between supply chain members, to enhance the economic value of PSFP contracts to suppliers, and promote positive employment and training outcomes, other actions are recommended. To reduce carbon emissions of PSFP, the recommended priority sequence is adjustment of waste disposal method, then menu composition, then transportation arrangements. Finally, the report summarises key findings of D6.2 on nutritional impacts of PSFP and the role of plate waste, and offers some integrated conclusions and recommendations based on both parts of WP6.

Appendix: Case studies description

Serbia

The dataset for the Serbian LOC and LOW cases comprised a selection of four schools, in each case, located either in the city of Belgrade or Novi Sad. Belgrade has a total population of 1.23million and a population density of 521 persons per km^2. There are 130 primary schools. Novi Sad is a city and municipality to northwest of Belgrade, with a population of 319,000, and a population density of 87 persons per km^2. There are 22 primary schools in Novi Sad city (37 in the municipality). All the schools, in both cases, are located in quite affluent districts, and Belgrade and Novi Sad are themselves more wealthy parts of Serbia.

Throughout Serbia, the provision of school meals is normally organised and managed at the individual school level, without any intervention from a central authority. All schools that provide all-day stay to pupils are required to provide meals (c.36% of all primary schools). There is no standard set price for school lunches, although a Strength2Food survey found the average price to parents nationwide is 143 RSD (€1.19, range €0.33–2.08). Lunches in Belgrade schools average 173 RSD (€1.45), while in Novi Sad the average price is 74 RSD (€0.62). (Novi Sad municipality has imposed a price freeze over the last 12 years, which explains why those lunch prices are very low). Around 75% of schools outsource catering to private firms, with the remainder preparing and cooking meals in-house and on site in school. Menus are designed by the catering firms or in-house cooks and typically comprise a single-option hot meal of soup and/or meat/fish dish with side vegetables and bread, plus dessert (fruit or cake/cookie). All the schools in both LOC and LOW cases undertake their catering in-house. LOC case schools only contract with 1–2 suppliers each, whereas LOW schools typically contract with 3–6. The sizes of the schools in both LOC and LOW cases are large (average pupil rolls = 932 and 1,176, respectively).

Italy: Parma case

The LOC-ORG case model is based in Parma, a wealthy municipality in Italy. The area has 33 primary schools in total, each with an average of 200 pupils, and an average meal uptake of >90%. All schools are obliged to offer meals, and the full price to parents is €6.18. Menus typically comprise a daily single-option hot main meal, comprised of cereals or starch-based first course (e.g. pasta, rice), a meat or fish based second course, side vegetables, bread and fruit. Desserts are served only on special occasions. A private catering firm prepares and cooks the meals off-site in a central kitchen, and then transports them to most schools in the municipality (a few schools have the ingredients delivered directly and cook on site). The staffing levels equate to 5–6 kitchen staff per school. The meals contract is renewed on a 6 year cycle, and the current catering firm, part of a large cooperative enterprise, has held it since 1995. The caterer subcontracts to 29 suppliers (of which ten are the main ones), and many of these are large-scale enterprises, with turnovers of >€100m. Although the PSFP contract in LOC-ORG case encourages local sourcing, the definition of local in the contract is broad, and there is no minimum threshold specified. 24 of the 29 suppliers are based >100 km from Parma. As will be seen, this has implications for the sustainability outcomes of LOC-ORG case. The five featured schools in the case are medium to large sized (average pupil roll = 371) and have high meal uptakes (71–95%). Several food and health-related initiatives have been organised for all schools in the municipality.

Italy: Lucca case

Lucca is also a relatively wealthy municipality in Italy. The area has 29 primary schools in total, each with an average of 100 pupils. All schools are obliged to offer meals, and the full price to parents is €5.00. Menus are designed according to the same guidelines as Parma municipality, and therefore comprise the same elements as LOC-ORG case, although dessert can be served more often (substituting for fruit). A private catering firm prepares and cooks all the meals in a central kitchen, and then transports them to school sites. The staffing levels equate to 3–4 staff per school. The meals contract is renewed on a 9-year cycle, and the current caterer, part of a regional corporate enterprise, has held it since 2002. The caterer subcontracts to nine suppliers, around half of which are large enterprises with turnovers >€100m. Around half of the suppliers are located inside the region. The five featured schools have an average pupil roll of 182, which means they are larger than the municipal average but smaller than the LOC-ORG schools. Meal uptake is very high in four schools (88–90%), and much lower in one school (46%). At least one major food-related educational project has been organised for schools in the municipality by the catering firm, as part of the specifications of the contract.

Note

1 www.sueatablelife.eu/it.html, www.strength2food.eu/.

Bibliography and references

Capone, R., Iannetta, M., El Bilali, H., Colonna, N., Debs, P., Dernini, S., Maiani, G., Intorre, F., Polito, A., Turrini, A., Cardone, G., Lorusso, F., & Belsanti, V. (2013). Preliminary Assessment of the Environmental Sustainability of the Current Italian Dietary Pattern: Water Footprint Related to Food Consumption. *Journal of Food and Nutrition Research*, Vol. 1, No. 4, 59–67. Retrieved from http://pubs.sciepub.com/jfnr/1/4/5

CREA. (2018). *Dossier Scientifico delle Linee Guida per una sana alimentazione* [Scientific Dossier of Guidelines for Healthy Eating]. Retrieved from www.crea.gov.it/web/alimenti-e-nutrizione/-/dossier-scientifico-linee-guida-per-una-sana-alimentazione-2018

Ferrari, M., Piccinelli, E., Martone, D., Catasta, G., & Turrini, A. (2023). Healthy Italian Dietary Pattern to Achieve 2030 Climate Target. *EC Nutrition* Vol. 18, No. 5, 22–37. Retrieved from www.researchgate.net/publication/372628459_Healthy_Italian_Dietary_Pattern_to_achieve_2030_climate_target

GHG Protocol. (2022). *Quantifying the Greenhouse Gas Emissions of Products PAS 2050 & the GHG Protocol Product Standard. A Short Guide to Their Purpose, Similarities and Differences*. Retrieved from https://ghgprotocol.org/sites/default/files/2022-12/GHG%20Protocol%20PAS%202050%20Factsheet.pdf

Hoekstra, A. K., & Chapagain, A. Y. (2004). *Water Footprints of Nations*. Retrieved from www.waterfootprint.org/resources/Report16Vol1.pdf

Lancaster, O., & Durie, S. (2008). *Technical Report. The Social Return on Investment of Food for Life School Meals in East Ayrshire*. Retrieved from https://socialvalueuk.org/reports/the-social-return-on-investment-of-food-for-life-school-meals-in-east-ayrshire/

Rossi, L., Berni Canani, S., Censi, L., Gennaro, L., Leclercq, C., Scognamiglio, U., Sette, S., & Ghiselli, A. (2022, March). The 2018 Revision of Italian Dietary Guidelines: Development Process, Novelties, Main Recommendations, and Policy Implications. *Frontiers in Nutrition*, Vol. 9. doi: 10.3389/fnut.2022.861526tks

Strength2Food. (2019, February). *Evaluation of Environmental, Economic and Social Impacts of Different Models of Public Sector Food Procurement in a School Context, Final Report*. Retrieved from www.strength2food.eu/wp-content/uploads/2019/03/D6.3-Evaluation-of-environmental-economic-and-social-impacts-of-different-PSFP-models-compressed.pdf

11 The European "meal voucher" market and its evolution

Maurizio Mariani

Introduction

In the European market, meal vouchers have long been an established means of facilitating meal benefits for employees. Their appeal in several member states, including France and Italy, lies in their tax efficiency, and convenience. These vouchers, while traditionally paper-based, have seen a digital transformation in recent years, adapting to the rapidly changing landscape of workplace benefits.

Italy has seen the emergence of a new business model during the COVID-19 pandemic called the "Mensa Diffusa" or distributed canteen. This decentralised system is different from the traditional meal vouchers market and allows employees to access meals from a network of local partner restaurants using a personalised app that also provides nutritional information. This innovative approach offers a wider range of food choices to cater to diverse preferences and provides a more inclusive and flexible lunch break experience while promoting community integration and the local economy.

The juxtaposition of the meal voucher system and Italy's distributed canteen model presents an intriguing exploration into the evolution of employee meal benefits. While the meal voucher system emphasises consistency and accessibility, the distributed canteen spotlights flexibility, diversity and community engagement. Italy's embrace of the latter hints at a potential shift in how businesses perceive and execute employee benefits, underlining the value of adaptability and local integration.

Yet, challenges exist. The harmonisation of these models across varied EU landscapes, preservation of nutritional standards and ensuring equitable access remain crucial. As the EU navigates the balance between standardisation and innovation, Italy's distributed canteen model may serve as a beacon for reimagining workplace meal benefits in a post-pandemic world.

The European meal vouchers market

Investigating into the public food service sector, we cannot ignore this market segment and his evolution, which, over the past 30 years, has effectively reduced meals served in public food services canteens in favour of commercial restaurants

DOI: 10.4324/9781003393023-13

and food shop by providing a "canteen replacement service" through the provision of meal vouchers, especially in small settings where fewer than 200 meals per day were being served.

To regulate the benefit, countries have varying laws specifying the maximum tax exemption for employers and maximum employee contribution, and also controlling the types of shops and restaurants allowed to participate in the scheme. Certain items, such as alcohol and tobacco, are not allowed and there is often a cap to the daily amount to be spent with vouchers. Vouchers were originally in paper form although they are now increasingly available as a chip and PIN card, or even completely dematerialised (QR code or e-platform).

As an alternative to the traditional management of a contract for onsite food service, contracting stations, whether public or private, have had a choice for years towards the so-called canteen replacement model.

We are talking about meal vouchers, created in the United Kingdom in 1946 and developed in France since 1967 thanks to a law that exempted sums paid under this title from taxes and social contributions. Since the 1970s, they have spread to some 40 countries around the world.

The European meal vouchers market is highly competitive and diverse, with various companies providing meal voucher services to businesses and individuals. Estimated to be worth over €16 billion annually, in France alone, the country that has developed this market more than others, the value is around 8 billion euros,[1] while in Italy it is 3.2 billion[2] and in Belgium 2.8 billion.[3] Then follows Spain with 0.6 billion,[4] while some Eastern European countries such as Romania, Slovakia or the Czech Republic are starting to move more and more important figures, the second world largest market is Brazil after France.

Europe's three largest meal voucher issuing companies are Edenred, Sodexo and Up Group, which hold over 90% of the market share.

We estimate that over 10 million people in Europe receive meal vouchers as a form of employee benefit or government assistance and more than 300.000 restaurants and food shops are part of the network accepting this kind of voucher instead of cash or credit cards.

The average value of a meal voucher in Europe is around €7.00 with most vouchers being valid for a single meal, and the use of digital meal vouchers is growing with an estimated 25% of all meal vouchers now being issued electronically by providing chip cards or apps to users.

Without questioning the economic and social value of this sector, the strong concentration on a few companies is evident, and the recent scandal in France and altercations in Italy leave quite a few perplexities about this "business model", which I now believe needs a new impetus, not only in relation to the simple switch from paper vouchers to digital vouchers that is now underway.

Unfortunately, the few publicly available data on this market refer to the broader "employee benefits solutions" sector, which is certainly fast-growing; estimated to date at around \$70 billion in Europe alone, and \$221 billion worldwide, and estimated to reach \$346 billion in 2030[5]

Recent scandals in France and altercations in Italy place this market under the spotlight

The French Competition Authority at the end of 2019 has imposed penalties totalling almost €415 million[6] on the four incumbent issuers of meal vouchers for cartel behaviour.

> The meal voucher sector is highly concentrated: four operators (Edenred France, UP, Natixis Intertitres and Sodexo Pass France) hold almost 100% of the meal voucher market (paper and dematerialised vouchers combined). The sector is also unique in that these four operators are brought together within a common structure, the Centrale de Règlement des Titres (CRT), which pools the processing of paper vouchers and their reimbursement to retailers. Other players issue only dematerialised vouchers, such as Moneo Payment Solutions (Moneo Resto), Caisse Fédérale du Crédit Mutuel (Monetico Resto) and Octoplus (Resto flash). Their combined market share does not exceed 1.5%.

The "Centrale de Règlement des Titres" (CRT), which managed the four main companies issuing meal vouchers, closed its doors on 28 February 2023.[7]

The Italian Central Public Purchasing Agency (CONSIP) – motivated by its interest in maximising savings on public procurements – in evaluating the economic offer, has made the criterion of the discount on the face value of meal vouchers preponderant, incentivising issuing companies to adopt aggressive commercial policies, with the effect of transferring – following the provisions of Article 144, paragraph 6, letter a) of the procurement code – the discount to the purchaser on the commissions charged to the contracted establishments (Restaurants).

Suffice it to say that in the last two CONSIP tenders, the operators (i.e. restaurants) themselves paying, respectively in the years 2018–2020, on average a commission of 19.8% (Tender BP 8) and 16.78% (Tender BP 9).

To be precise, it should be noted that Article 144, para. 6 of Legislative Decree No. 50/2016 provides for the award of this kind of service exclusively on the basis of the criterion of the economically most advantageous offer, identified on the basis of the best quality/price ratio.

Consip is used to award a maximum of 85 points to the technical offer and 15 to the economic offer, however, the services requested for the first component can easily be ensured by all bidders, with the consequence "*de facto*", that the entire competition is focused on the element of the economic offer to the detriment of the qualitative one.

In response to this situation that exploded in the spring 2022, there were many protests from restaurant owners and restaurant associations, which met at a table during the month of May asking themselves and to CONSIP: why do Italian companies in the food chain join the meal voucher network despite the excessive fees charged to them?

Because the Italian market for meal vouchers (but we should assume it for the entire EU restaurants market) – with an annual value of EUR 3.2 billion, between the public (1.2 billion) and private sector (2.0 billion)- involves 2.8 million workers[8] who are customers of restaurants not only when using meal vouchers but also on other occasions of consumption.

For an individual restaurant or food shop, refusing meal vouchers can result in the ultimate loss of a customer.

In other cases, then, a dependency mechanism is created that forces meal vouchers to be accepted even when they are not economically viable. These are those activities mostly located close to workplaces with many employees, for which accepting food vouchers, even at a loss, serves to guarantee at least partial coverage of costs.

Facing this situation, which I define as the "financialization of the meal", the merchant, i.e. those who provide the meal service, are severely penalised by the very high commissions and management costs that can sum up to 30%[9] of the face value of the meal voucher received, this is particularly so with the still predominant use of paper voucher or chip cards.

An emerging business model: The case of the Italian "mensa diffusa"

We all know that digitalisation has played a major role in recent years, and has affected every aspect of our lives, including the vast and complex world of catering: digital menus, contactless payments and a boom in delivery services.

The employers lunch break has been no different. So over time it has evolved, changed and updated, both in terms of regulations and the tools available to employers and restaurants. Among these, the "*mensa diffusa*" or distributed canteen could prove to be a good opportunity.

Due to the increasing digitalisation and the development of smart working, which have been further accentuated after the COVID-19 pandemic, the existing system has been undermined. As a result, digital start-ups have emerged, which have developed apps that allow companies to create a network of affiliated restaurants. This has given rise to the phenomenon of "mensa diffusa" or *distributed canteen*, similar to what has already happened in the hotel sector in some old Italian villages, where individual houses have become an "albergo diffuso" or distributed hotel, by connecting them to a single digital management and marketing system.

The distributed canteen is a system that enables employees to have a meal provided by their employers at their preferred restaurant, just as if they were having lunch at their workplace canteen. This system is similar to the company canteen service or to the meal vouchers, but it operates through a network of local affiliated restaurants and shops. Employees can even recommend new establishments to join the network. Payments can be made via an app without the need for cards or cash and commission are largely inferior to those of meal vouchers, on average around 6% without any other cost related.

By using this system, employees can enjoy their meals wherever they are, even if they are working remotely, by paying with their smartphones. By utilising digital food vouchers, individuals can enhance their professional and social status, as well as develop a greater familiarity with digital technology.

Using dedicated applications (APPs) for managing corporate lunches can be a more efficient and sustainable alternative to traditional meal vouchers. These apps can help employers promote healthy eating habits while facilitating communication and enhancing employee welfare programmes, both in public and private organisations.

Leaving aside the tax/fiscal aspects of this new system, which vary, as with classic meal vouchers from country to country, and which are certainly not the obstacle to the spread of this new system, let us see what practical advantages it offers to organisations (employers) that decide to switch to this new system.

Meals are not cumulative, this means that each day the employee is entitled to a meal which, if not consumed, generates cashback for the employers; and even the daily amount not spent can be recovered by the employers through cashback (e.g. if the daily amount available is €10 but the employee only spends €8, the other €2 is returned to the employers)

This results in significant savings because, unlike meal vouchers,[10] it allows the employers to recover not only the meals not consumed but also the amount left over from each individual meal paid less than the daily amount set for the lunch.

With the distributed canteen, no counting of the following month's attendance is necessary, as is the case with meal vouchers, because by choosing this system, the counting is done automatically when the employee buys his meal, facilitating the payroll process.

In short, the distributed canteen can really represent, especially in these times, a cheaper and employee-friendly alternative to the company and institutions canteen and traditional meal vouchers.

How this sector can contribute to the development of a more sustainable and healthier local food service market

The concept of a distributed canteen is a remarkable innovation that stands as a powerful lever for sustainability and social change, capable of reshaping community lunch break dynamics and fostering environmentally friendly practices.

By decentralising the provision of catering services and integrating them within various local establishments, this model addresses a myriad of social and environmental challenges, contributing significantly to sustainable development and enhanced community cohesion.

One of the key sustainability levers of the distributed canteen model is its ability to support the local economy. It creates a network that connects local restaurants and eateries with a wider consumer base, thereby channelling more business to these establishments. This symbiotic relationship not only sustains local businesses but also ensures that money circulates within the community, reinforcing the local economy and make it easy to foster also circular economy systems.

Moreover, the distributed canteen model plays a pivotal role in reducing environmental footprint. Traditional canteens often require the construction and operation of dedicated facilities, which can have a substantial environmental impact. By utilising existing local establishments, the distributed canteen model circumvents the need for additional infrastructure, thereby minimising resource utilisation and overall environmental degradation.

Further, the collaboration between the distributed canteen and local businesses should provide an excellent platform to promote sustainable and healthy practices. These include sourcing local and seasonal ingredients, minimising food waste and utilising eco-friendly packaging and utensils. The cumulative impact of such practices can significantly contribute to environmental conservation and promote a culture of sustainability among consumers and businesses alike.

On the social change front, the distributed canteen model is a catalyst for enhancing social inclusivity. By offering access to a diverse array of dining options in different locations, it makes catering services more accessible to a wider population. This is particularly impactful in areas where constructing a traditional canteen is impractical or economically unviable, thereby addressing issues related to food accessibility and reducing social inequalities.

Furthermore, the distributed canteen should foster community cohesion by integrating a variety of eateries into its network. This will encourage individuals to explore different culinary cultures and establishments within their community, fostering interaction among community members, breaking down social barriers and nurturing a sense of unity and mutual understanding.

Empowering individuals with choices is another significant aspect of the distributed canteen model. The flexibility and variety inherent to this system allow individuals greater autonomy in their dining choices, enhancing their overall experience and encouraging a sense of responsibility and mindfulness regarding sustainable and ethical consumption.

In conclusion, the distributed canteen model emerges as a multifaceted solution intertwining economic, environmental and social dimensions. It reimagines the approach to lunch break services, contributing to building resilient local economies, fostering environmental stewardship, enhancing social inclusivity and nurturing community ties. The embodiment of this innovative model signifies a progressive step in aligning catering services with the broader goals of sustainable development and social equity, marking a paradigm shift in community lunch break.

The distributed canteen and meal vouchers serve as two distinctive strategies to cater to the meal provision needs of employees and occasionally, students. Each strategy holds its unique set of advantages and disadvantages and can be applied either complementarily or alternatively, contingent upon the distinct requirements of a community or organisation.

However, the distributed canteen is not devoid of challenges. It might necessitate more intricate management and logistics to coordinate and oversee the various collaborations with commercial establishments. Another aspect to consider is the variability in cost, as the price per meal can fluctuate depending on the chosen restaurant.

On the other hand, meal vouchers have emerged as a convenient solution for employers and less for restaurant owners. They are straightforward to distribute and manage, making them an ideal choice for companies seeking practicality. Their wide acceptance across a diverse range of commercial establishments bestows users with flexibility, enhancing their lunch break experience. A pivotal advantage of meal vouchers lies in cost control, as the fixed value of each voucher enables organisations to manage their budgets effectively.

However, meal vouchers have some limitations. The number of lunch break establishments that accept meal vouchers may be limited due to the high commission fees and, in some cases, restaurant payments are delayed for several months. This can pose a risk to the restaurant, as was the case in Italy where the issuer went bankrupt, causing a dramatic economic impact on the local restaurants.[11];

In conclusion, the decision between implementing a distributed canteen or opting for meal vouchers is contingent upon the individual needs and priorities of an organisation or entity. A tailored approach, potentially incorporating a blend of both strategies, could be the key to devising a balanced solution that amplifies the benefits while mitigating the disadvantages. The exploration of these two models sheds light on the diverse avenues available for addressing meal provision needs, each with its unique characteristics and implications.

New opportunities for local authorities?

Local authorities already engaged in serving healthy & sustainable meals in the school canteen should create a whole local food service system, enlarging the system itself to the distributed canteen through a quality label/scheme gathering local restaurants, where public employees should get their meals using the virtual voucher (APPs). Managing it directly or in a joint venture with private operators, (PPP) but leading this process of change, instead of managing a public tender to assign the whole service, make it easier and select the restaurant by a kind of self-certification process, verified by the consumers themselves via unique APPs.

A sustainable & healthy quality label/scheme for local restaurants would be a great way to promote environmentally and socially responsible practices within the food industry. Such a label could be used to identify restaurants that source their ingredients locally, use sustainable farming practices, reduce waste and minimise their carbon footprint.

The label could be based on a set of criteria that includes factors such as using locally-sourced ingredients, organic or sustainably grown produce, reducing food waste, energy efficiency and use of renewable energy sources. The criteria could also consider the restaurant's social and ethical practices, such as fair labour practices, community engagement and support for local farmers and producers.

By promoting sustainable practices, such a scheme could encourage local restaurants to adopt more environmentally responsible practices while also providing customers with a straightforward way to identify restaurants that share their values. Additionally, the label could promote local economies and support local farmers and producers, which is vital for building resilient and sustainable food systems.

In conclusion, a sustainable and healthy quality scheme for local restaurants could be a powerful tool for promoting sustainable and socially responsible practices within the food industry while also providing customers with a straightforward way to identify environmentally conscious restaurants fostering local economy, and certainly increase the Social Return on Investment ratio.

It is yet to be determined if this system can be operated in-house and thus be exempt from the competitive procurement process which requires a public tender. However, it will be interesting to observe how new players such as "satispay", "swile" and "coverflex" will affect this market from the perspective of those providing the service. Undoubtedly, the competitiveness of the meal voucher market will increase, and it is possible that there will be a reduction in commissions for restaurants.

Notes

1 www.cntr.fr/chiffres/.
2 www.slideshare.net/Fipe-Confcommercio/il-mercato-dei-buoni-pasto-227267721.
3 www.viabelgium.be/titre-repas/index.html.
4 www.expansion.com/fiscal/2022/12/02/638514ff468aebb47e8b457a.html.
5 www.fortunebusinessinsights.com/meal-voucher-and-employee-benefit-solutions-market-104412.
6 (www.autoritedelaconcurrence.fr/fr/communiques-de-presse/lautorite-de-la-concurrence-inflige-des-sanctions-dun-total-de-pres-de-415).
7 https://partenaire.edenred.fr/s/article-details?id=ka51i000000PL3EAAW.
8 www.slideshare.net/Fipe-Confcommercio/il-mercato-dei-buoni-pasto-227267721.
9 www.federdistribuzione.it/app/uploads/2021/07/IL-PRANZO-FUORI-CASA_2020_DEF.pdf.
10 www.capital.fr/votre-argent/lincroyable-somme-perdue-chaque-annee-par-les-francais-sur-leurs-tickets-restaurant-1456302.
11 www.ansa.it/liguria/notizie/2022/10/07/fallimento-quigroup-chiuse-indagini-per-31-persone_2a6176df-b7c0-4617-93da-aa0b1d88efaa.html.

Part II

Future scenarios for sustainable procurement of food and catering

12 The impact of the "Farm to Fork" strategy on sustainable food and catering procurement

Roberto Caranta

Introduction

Public procurement regulation in the EU has traditionally served the aim of opening domestic markets to EU-wide competition. Since 1971, secondary EU (then EEC) law rules have provided for open and transparent contract award procedures geared at tearing down protectionist barriers. EU public procurement (and concessions, but the distinction is immaterial here) rules thus regulate "how to buy". It is for the Member States and their contracting authorities to decide "what to buy", provided that in their choices they do not infringe the general EU Treaties provisions on the internal market and specific EU secondary rules.

The general EU law principle of non-discrimination and the strongly embedded internal market 'orthodoxy' have been seen as a limit to the inclusion of social and environmental considerations in procurement choices. In the past three decades however, thanks to the decisive push of the Court of Justice and more recently of the European Parliament, environmental and social considerations have slowly infiltrated the EU public procurement regulatory framework, widening the margins of choice left to the Member States and to their contracting authorities. Still, this cannot translate into reinstituting trade barriers (see the contributions collected by Sjåfjell & Wiesbrock (eds), 2016; see also Dragos & Neamtu, 2014).

This hands-off approach has been turned on its head by and following the European Green Deal (European Commission, 2019a). Faced with the need to change linear production and consumption models to address immense societal challenges such as global warming and climate change and to contribute to achieving the SDGs, the European Commission has tabled, and the European Parliament and the Council are approving a raft of legislative proposals making sustainable public procurement (henceforth SPP) mandatory in a growing number of sectors. In this way, EU law moves from regulating just 'how to buy' into giving the Member States and their contracting authorities rules about 'what to buy' (Janssen & Caranta, 2023).

Following the Farm to Fork Strategy implementing the European Green Deal, food and catering procurement are expected to follow the same path towards mandatory SPP, and a so far delayed Commission's proposal is expected. This policy evolution is part of a global phenomenon. "Over the past decade, national, regional and local governments in various low-to high-income countries have been

DOI: 10.4324/9781003393023-15

developing public food procurement (PFP) initiatives that use public purchasing power and a regular demand for food as a policy instrument to promote sustainable development" (FAO, 2021).

In this chapter, after sketching the current EU SPP legislative framework (§ 2), and the European Green Deal (§ 3), the Farm to Fork Strategy will be introduced (§ 4) to assess the potential for mandatory SPP of food and catering services within but also outside the Farm to Fork Strategy (§ 5). The conclusions will indicate the legal basis for the current shift towards mandatory SPP and point to how sustainable considerations might otherwise be lawfully included in food procurement strategies (§ 6). For want of space, we will not go here into a detailed analysis of what makes 'sustainable' food (see Schebesta & Plana Casado, 2023, pp. 133 ff).

General EU SPP legal framework

To make a long story short, under the currently in force 2014 EU procurement directives, SPP is a possibility for contracting authorities (detailed analysis in Sjåfjell & Wiesbrock (eds), 2016 and Semple, 2015). They have also introduced innovative tools, such as life-cycle costing. Most of the SPP provisions in the 2014 directives however simply enable national contracting authorities and entities to pursue SPP. Moreover, the possibility to make SPP choices is not without limits. The general principles of EU law, such as non-discrimination and transparency shall still be complied with and – to avoid greenwashing – sustainability requirements must be capable of objective verification. With a provision that mutatis mutandis may be applied to other steps in the award procedure, the first part of Article 67(4) of Directive 2014/24/EU provides that

> [a]ward criteria shall not have the effect of conferring an unrestricted freedom of choice on the contracting authority. They shall ensure the possibility of effective competition and shall be accompanied by specifications that allow the information provided by the tenderers to be effectively verified in order to assess how well the tenders meet the award criteria.

Moreover, any SPP requirement must be linked to the subject matter of the contract. Under Article 67(3), a requirement "shall be considered to be linked to the subject matter of the public contract where they relate to the works, supplies or services to be provided under that contract in any respect and at any stage of their life cycle". This wide provision considerably waters down the link to the subject matter requirement and generously enables the insertion of SPP considerations in purchasing decisions, with the exclusion set in the last phrase of Recital 97 of conditions relating to "general corporate policy, which cannot be considered as a factor characterising the specific process of production or provision of the purchased works, supplies or services" (see Semple, 2021).

A principle of sustainability was enshrined in Article 18(2) of Directive 2014/24/EU. In TIM the Court of Justice held that "that such a requirement constitutes, in the general scheme of that directive, a cardinal value with which the Member States must ensure compliance" (paragraph 38) (Andhov, 2021; European Court of Justice,

2020). Still, only in specific sectors EU has mandated SPP. For instance, Regulation (EC) No 106/2008, the so-called EU Energy Star Regulation requires contracting authorities to introduce in their public contracts a certain level of energy efficiency.

The European Commission provides guidance to public buyers on how to use SPP in the form of guides and SPP criteria. For instance, in 2016, to coincide with the deadline for the transposition of the 2014 procurement and concessions directives in the Member States, the Commission published the 3rd edition of the *Buying Green! – A Handbook on green public procurement* (European Commission, 2016). The Handbook is the Commission's main guidance document to help public authorities buy goods and services with a lower environmental impact and provides sector-specific green public procurement (henceforth GPP) approaches, including for food and catering services.

Alongside with the different guidance documents, the Commission also publishes good practice examples and very detailed GPP criteria, accompanied by a technical background report. GPP criteria are articulated in 'core criteria' that are suitable for use even by small contracting authorities, address the key environmental impacts and are designed to be used with minimum additional verification effort or cost increases, and 'comprehensive criteria' referring to the best environmental products available on the market. Again it must be stressed that the use of those criteria is voluntary for contracting authorities in the Member States. Basically, they are soft law in nature (Schebesta & Plana Casado, 2023, p. 131).

In this context, in 2019 the Commission published a 52-pages long new GPP criteria for food, catering services and vending machines. The key environmental impacts of food procurement associated with the food's production identified are "the combustion of fossil fuels and energy use for different activities, land use or land-use change and water use and water pollution". Further environmental impacts are associated with specific food product categories, such as the production and use of pesticides and chemical fertilisers, soil degradation, emissions of methane and nitrate etc. In catering services, "energy and water use are important contributors to the overall environmental impact, as are waste generation and waste management" (European Commission, 2019b, p. 6).

The 2019 GPP criteria for food, catering services and vending machines have limited scope as they include neither social nor health/nutritional considerations (Schebesta & Plana Casado, 2023, pp. 131 ff.). It is also true that the same European Pillar of Social Rights were not overly ambitious concerning public procurement. The 2021 European Pillar of Social Rights Action Plan simply foresaw the updating of the Procurement of Innovation and the Buying Social guides and required contracting authorities in the Member States to increase the uptake of socially responsible public procurement (SRPP) criteria "in line with the Commission's forthcoming guidance notices" (European Commission, 2021b, p. 28).

The European Green Deal

The tepid acceptance of SPP by EU institutions – the Commission first among them – has been much changed in the past few years following the European Green Deal (Tátrai & Diófási-Kovács, 2021). The Communication of the EU Green Deal

(2019a) is not light on ambitions, the mandate being to transform the EU economy for a sustainable future by designing a set of deeply transformative policies. According to the Commission,

> To deliver the European Green Deal, there is a need to rethink policies for clean energy supply across the economy, industry, production and consumption, large-scale infrastructure, transport, *food and agriculture*, construction, taxation and social benefits. To achieve these aims, it is essential to increase the value given to protecting and restoring natural ecosystems, to the sustainable use of resources and to improving human health. This is where transformational change is most needed and potentially most beneficial for the EU economy, society and natural environment.
>
> (European Commission, 2019a, p. 4, *emphasis added*)

Food and agriculture are bound to be at the centre of change if the European Green Deal is to deliver. The Communication acknowledges that

> From 1970 to 2017, the annual global extraction of materials tripled and it continues to grow, posing a major global risk. About half of total greenhouse gas emissions and more than 90% of biodiversity loss and water stress come from resource extraction and processing of materials, fuels and *food.*
>
> (*ibid.*, p. 7, *emphasis added*)

Tools must meet ambitions and, given that public procurement commands a sizeable portion of the GDP in the EU, it must obviously be enlisted to the wider effort towards sustainability. The Communication is content with stating that "[p]ublic authorities, including the EU institutions, should lead by example and ensure that their procurement is green. The Commission will propose further legislation and guidance on green public purchasing" (ibid., p. 8). The new Circular Economy Action Plan (CEAP) adopted in March 2020, one of the main building blocks of the European Green Deal, went further, announcing the proposed introduction of mandatory SPP criteria (European Commission, 2020b). Since "[p]ublic authorities' purchasing power represents 14% of EU GDP", it "can serve as a powerful driver of the demand for sustainable products" (ibid., p. 5). However, experience has shown that instruments such as the EU GPP criteria "have reduced impact due to the limitations of voluntary approaches" (ibid., p. 3). To instead tap into the potential offered by public procurement, the Commission "will propose minimum mandatory green public procurement (GPP) criteria and targets in sectoral legislation and phase in compulsory reporting to monitor the uptake of Green Public Procurement (GPP)" of course "without creating an unjustified administrative burden for the public buyers" (ibid., p. 5). On the same beat, the EU Green Deal Investment Plan (2020a) stated:

> The Commission will propose minimum mandatory green criteria or targets for public procurements in sectoral initiatives, EU funding or product-specific legislation. Such minimum criteria will "*de facto*" set a common definition

> of what a "green purchase" is, allowing the collection of comparable data from public buyers and setting the basis for assessing the impact of green public procurements. Public authorities across Europe will be encouraged to integrate green criteria and use labels in their procurements (European Commission, 2020a, p. 12).

The shift towards mandatory criteria is in line with the JRC recommendations in its *Concepts for a sustainable EU food system* (Bock et al., 2022). The report stressed that:

> While voluntary measures and agreements. . . might be useful to initiate change in the short term, substantial change requires the formulation of ambitious and effective binding rules. Such rules would provide the necessary reliability and predictability for businesses by setting ambitious goals in combination with a practical timeframe.
>
> (*ibid.*, pp. 7, 58)

The recommendations go well beyond public procurement as they concern the EU food system generally, but the reasoning fits to public procurement as well.

A huge number of proposals have been tabled by the Commission, covering batteries, construction products, energy etc (see the contributions in Janssen & Caranta, 2023). Food and catering procurement is expected to follow soon under the Farm to Fork Strategy.

The Farm to Fork Strategy

As already indicated, food systems have their share of responsibility for the challenges our societies are facing (for some data and references see European Commission, 2020c, p. 6 ff.). Food is at the core of the European Green Deal and one of the lines of action foreseen by the Commission goes by the euphonic words of Farm to Fork. The Commission wants to design a "fair, healthy and environmentally-friendly food system". To this end, it announced the presentation of the 'Farm to Fork' Strategy to strengthen efforts to "tackle climate change, protect the environment and preserve biodiversity" (European Commission, 2019a, p. 12). The Farm to Fork Strategy is further expected to "contribute to achieving a circular economy. It will aim to reduce the environmental impact of the food processing and retail sectors by taking action on transport, storage, packaging and food waste" (*ibid.*, p. 12).

The Strategy acknowledges that

> People pay increasing attention to environmental, health, social and ethical issues and they seek value in food more than ever before. Even as societies become more urbanised, they want to feel closer to their food. They want food that is fresh, less processed and sustainably sourced.
>
> (European Commission, 2020c, p. 2)

The Farm to Fork Strategy was conditioned by and took stock from the COVID-19 pandemics raging when it was published and introduced 'resilience' as a relevant consideration if not a new policy objective: "The COVID-19 pandemic has underlined the importance of a robust and resilient food system that functions in all circumstances". Unsurprisingly, "calls for shorter supply chains have intensified during the current outbreak" (European Commission. 2020c, p. 2). Therefore, "with a view to enhance resilience of regional and local food systems, the Commission in order to create shorter supply chains will support reducing dependence on long-haul transportation" (*ibid.*, p. 12, where an indication of the staggering weight of food and forestry products transported yearly; see also Bock et al., 2022, p. 9).

Mandatory SPP of food and catering services

In the Farm to Fork Strategy Communication (2020c), SPP features as one of the tools to increase the share of organic farming. It is first referred to in connection to organic farming which "has a positive impact on biodiversity, it creates jobs and attracts young farmers" (p. 8). The EU is thus expected to take a number of measures, first of all in the framework of the Common Agricultural Policy (CPA) and of the Common Fisheries Policy (CFP). The Commission announced an Action Plan on organic farming expected to "boost demand through promotion campaigns and green public procurement". The combined policy initiatives should help in reaching "the objective of at least 25% of the EU's agricultural land under organic farming by 2030 and a significant increase in organic aquaculture" (*ibid.*, p. 8). More generally, "[t]o improve the availability and price of sustainable food and to promote healthy and sustainable diets in institutional catering, the Commission will determine the best way of setting minimum mandatory criteria for sustainable food procurement". Besides boosting sustainable farming systems, such as the already referred to organic farming, mandatory SPP "will help cities, regions and public authorities to play their part by sourcing sustainable food for schools, hospitals and public institutions" (*ibid.*, p. 13; see also European Commission, 2021a, p. 3). The Commission itself will 'lead by example' and "reinforce sustainability standards in the catering contract for its canteens" and will also strengthen the sustainability of the EU school scheme (European Commission, 2020c, p. 13) (see Schebesta & Plana Casado, 2023).

Action 22 of the Draft Action Plan included in the Annex to the Farm to Fork Strategy requires the determination of "the best modalities for setting minimum mandatory criteria for sustainable food procurement to promote healthy and sustainable diets, including organic products, in schools and public institutions". The Action plan also foresees a more general proposal for a legislative framework for sustainable food systems.

The 2021 Commission Communication on An Action Plan for the development of organic production aims to include organic production within GPP (European Commission, 2021a, p. 2). A section of the Communication (1.2.) is dedicated to "Promoting organic canteens and increasing the use of green public procurement". The Commission recalls that the GPP criteria for food, catering services and vending machines were reviewed in 2019, but laments that there is still "a lack of knowledge among public administrations – particularly local ones – on the possibilities offered by GPP". The Farm to Fork Strategy is then referred to as to the need to determine how best to set minimum mandatory criteria (*ibid.*, pp. 7–8).

More specifically, Action 3 of the Action Plan for the development of organic production, requires the Commission to contribute to boosting the awareness of the 2019 GPP criteria and to "integrate organic products into the minimum mandatory criteria for sustainable food public procurement to be developed as part of the legislative framework for sustainable food systems by 2023". Moreover, "the Commission will use the national action plans on organic farming to monitor the application of GPP and call on Member States for an increase in the use of GPP by public authorities" (*ibid.*, p. 8). As Schebesta and Plana Casado (2023, p. 128) indicate, the Action Plan shifts the emphasis from GPP to SPP and mandatory criteria are expected to be part of the forthcoming Framework for a Sustainable Food System (FSFS).

Additionally, section 1.3. of the Action Plan for the development of organic production (Reinforcing organic school schemes) indicates that the Member States "should prioritise the distribution of organic products under the EU school scheme, through selection or award criteria in procurement procedures and/or through more favourable conditions" (European Commission, 2021a, p. 8).

The possibility of adopting mandatory requirements for sustainable food procurement is now discussed in connection with the proposed Framework for a Sustainable Food System (FSFS, expected by the end of 2023), which is also to cover the sustainability labelling of food. The scope of this proposal thus goes beyond 'green' public procurement, and targets 'sustainable' public procurement instead (see Schebesta & Plana Casado, 2023).

The forthcoming FSFS proposal is therefore expected to include mandatory sustainable food procurement provisions, but it is yet unclear what will be the 'best modalities' for this. A recent research has highlighted the great variety of possible approaches to mandatory SPP. Summarising its result, Willem Janssen distinguishes between (1) minimum mandatory public procurement requirements; (2) target; or (3) product-specific legislation (2023, p. 17). Since, however, the latter limit the possibility to place on the market products or services not meeting given requirements, it is not specific to public procurement so that we can focus on the alternative between minimum requirements and targets. Concerning the former, it is also to be distinguished between (1a) substantive criteria, setting for instance minimum quality thresholds (technical specifications) or prizing better quality (award criteria), and (1b) procedural criteria, e.g. requiring contracting authorities to motivate the choice for the lowest price as an award criterion (*ibid.* p. 18).

The different approaches to mandatory SPP each have pros and cons and this quite independently from the good or service purchased. Substantive minimum criteria are easy to apply for contracting authorities but take away all freedom of choice from them and might either be too strict for some operating in markets where sustainability is yet to develop or too light but still constraining for more ambitious contracting authorities. Arguably, however, the latter issue might be much mitigated by allowing willing contracting authorities to exceed mandatory requirements to be read as a floor, not as a ceiling. Depending on the possibly limited intensity of judicial review in any given Member State, procedural requirements might turn in a box ticking exercise without any real impact on the sustainability of buying choices. Targets are difficult to measure, and somehow recall Soviet era five-year planning (a more detailed analysis in Caranta & Janssen, 2023, pp. 257 ff.).

Concerning specifically food and catering services, Schebesta and Plana Casado foresee a gradual "phasing in" of sustainable food public procurement obligations becoming gradually more restraining over time. The difficulties in defining what sustainable foods are favours "setting targets rather than mandating specific contract clauses. This leaves some discretionary policy space for the prioritisation of policy and the importance of food sustainability dimensions to be done at the local level" (2023, p. 145).

In my view, however, such an approach will make even more difficult to check whether and to what extent contracting authorities are indeed pursuing sustainability goals. It seems therefore preferable to start with minimum mandatory substantive requirements based on the already existing GPP criteria and then expand them gradually to incorporate further aspects of sustainable procurement, including social ones with successive revisions of the criteria. The JRC *Concepts for a sustainable EU food system* highlights both the complexity of the multifaceted notion of sustainability and the need to develop evidence-based frameworks to assess it (Bock et al., 2022, pp. 11 and 28 ff respectively).

Sustainability criteria in food and catering procurement – be they mandatory or designed by a contracting authority, but the same is true for all procurement – need to be 'measurable and monitored' (Bock et al., 2022, p. 49) to avoid discrimination and unlawful limitations of the competition on the lines e.g. of Article 67(4) of Directive 2014/24/EU. This again stresses the need for a "sustainability assessment framework" (*ibid.* p. 56). This, however, does not mean that the framework cannot grow incrementally by adding further elements of sustainability as the data required and the assessment capacity grows, including in contracting authorities. Simplification and streamlining of the present labels babel will also help (*ibid.*, p. 34; see also Caranta, 2016, pp. 111 ff).

One more reason to prefer minimum mandatory substantive requirements is that their breach, unlike that of targets and procedural requirements, is easily justiciable even in those jurisdictions where judicial review is hands-off (Lichère & Sulpice, 2023, pp. 237 ff.). And reliance on enforcement by economic operators is at the core of EU public procurement law and of EU law more generally (Caranta, 2019).

An example of a mandatory criterion is to be found in the recent legislation to fight deforestation (European Union, 2023). Already in the Farm to Fork Strategy it was anticipated that, "[t]o reduce the EU's contribution to global deforestation and forest degradation, the Commission will present in 2021 a legislative proposal and other measures to avoid or minimise the placing of products associated with deforestation or forest degradation on the EU market" (European Commission, 2020c,

p. 17). Under Article of 25 of Regulation (EU) No 1115/2023 Member States shall provide for penalties in case of infringements of the Regulation by operators and traders, including, under lett. (d) "temporary exclusion for a maximum period of 12 months from public procurement processes and from access to public funding, including tendering procedures, grants and concessions". Sustainability concerns today move EU lawmakers towards also regulating aspects that might be labelled as "whom not to buy from".

Concerning specifically social aspects, while the European Pillar of Social Rights does not push the matter far, the eventual approval of the Corporate Sustainability Due Diligence Directive might signal an important change (European Commission, 2022). While the proposal does not itself refer to public procurement, the new directive will introduce social (and environmental) obligations that the Member States are to act upon under Article 18(2) of Directive 2014/24/EU (Schoenmaeker, 2023; Uysal, 2021).

Conclusion

The shift towards "what to buy" in EU public procurement is momentous. In a multilevel polity, Brussels needs to prove its competence at every step inevitably and this raises the question of the legal bases. A first indication comes from the first two phrases in Article 3(3) of the Treaty on European Union (TEU) laying down the aims the EU is to pursue, according to which

> The Union shall establish an internal market. It shall work for the sustainable development of Europe based on balanced economic growth and price stability, a highly competitive social market economy, aiming at full employment and social progress, and a high level of protection and improvement of the quality of the environment. It shall promote scientific and technological advance. It shall combat social exclusion and discrimination, and shall promote social justice and protection, equality between women and men, solidarity between generations and protection of the rights of the child.
>
> (European Union, 2016)

The three components of sustainable (economic, social and environmental) are all covered in the provision. While Article 3(3) is very much looking inward the external dimension is reflected in Article 3(5) calling the Union to

> uphold and promote its values and interests and contribute to. . . the sustainable development of the Earth, solidarity and mutual respect among peoples, free and fair trade, eradication of poverty and the protection of human rights, in particular the rights of the child, as well as to the strict observance and the development of international law, including respect for the principles of the United Nations Charter" (see also Article 21(2)(e) and (f), recalling the pillars of sustainable development with reference to the EU external action).

This, however, is just the start, the aims of the EU, not its competences. Concerning the latter, Article 3(6) TEU is circumspect with overkill by adjectives as it provides that the EU "shall pursue its objectives by appropriate means commensurate with the competences which are conferred upon it in the Treaties" (European Union, 2016). The principles of conferral, subsidiarity and proportionality in Article 5 TEU potentially further restrain the role of the EU. For the competences we need to look into the Treaty on the Functioning of the European Union (TFEU) (European Union, 2012). The only EU exclusive competence incidentally touching on environmental sustainability concerns the conservation of marine biological resources, but within the common fisheries policy, while also social aspects might be involved in the common commercial policy (Article 3(1)(d) and (e) TFEU) (Marín Durán, 2020). Sustainability is instead potentially involved in many shared competences listed in Article 4(2) TFEU. Shared competences are however defined therein following two different patterns, with or without the caption "for the aspects defined in this Treaty". Present the caption, the proper definition of the competence of the EU depends on other, more specific provisions in the TFEU, so that the competence is further restricted. In this category falls 'social policy' (lit. (b)); EU competencies in this area are defined in Articles 151 ff TFEU and normally do not go beyond coordinating and encouraging the Member States. Among the shared competences relevant here are those concerning the internal market (lit. (a)); agriculture and fisheries (excluding the conservation of marine biological resources: lit. (d)); the environment (lit. (e)); consumer protection (lit. (f)) and energy (lit. (i)). Identifying a EU competence which is at least shared leads to the finding of a possible legal base for EU legislation as under Article 2(2) TFEU, when a competence shared between the EU and the Member States, the Union and the Member States may legislate and adopt legally binding acts in that area, but the Member States may exercise their competence only to the extent that the Union has not exercised its own.

Coming back to public procurement and concession law, clearly the EU has competence to introduce legislation on all aspects relating to GPP following the appropriate procedure among those laid down under Article 192 TFEU. This legal basis may or may not be coordinated with Article 114 TFEU, detailing the competence of the EU to harmonise legislative provisions concerning the internal market. The two may be combined and under Article 114(3), legislative proposals "concerning health, safety, environmental protection and consumer protection, will take as a base a high level of protection". The legal basis for clauses concerning human and social rights is less clear-cut, as in this area there is no correspondence to Article 192 TFEU. Article 114 remains the default option, and legislation based on that provision will have to integrate the social goals listed in Article 8 to 10 TFEU. Other legal bases might be relevant. The proposal for a Corporate Sustainability Due Diligence Directive cleverly combines Articles 50 and 114 TFEU as a legal base, the former concerning freedom of establishment (European Commission, 2022).

The reason for involving Article 114 as a legal basis for mandatory sustainable requirements is that lacking action at EU level the Member States and their contracting authorities will each go down their own way towards SPP. Italy is a great case in point, with minimum mandatory SPP criteria for food and catering (Botta, 2023). However, this risks leading to a high fragmentation of the internal markets according to more or less demanding requirements that could even disguise local preferences while not contributing into achieving the objectives set in the European Green Deal. The criteria used by Member States should be similar to avoid a distortion of the single market and a reduction of EU-wide competition. Moreover, as the Commission avows in introducing the GPP criteria on its website,

> [h]aving common criteria reduces considerably the administrative burden for economic operators and for public administrations implementing GPP. Common GPP criteria are of a particular benefit to companies operating in more than one Member State as well as SMEs (whose capacity to master differing procurement procedures is limited).
>
> (European Commission, 2008)

This is obviously even truer when considering mandatory SPP criteria.

Waiting for EU mandatory SPP criteria, and hoping them to be minimum criteria for the reasons explained here, contracting authorities maintain – and will maintain some room for manoeuvre if the criteria enacted are indeed minimum – their ability to choose more sustainable options under the general public procurement and concession rules.

Indeed, EU law leaves intact and is expected to continue leaving intact the freedom of contracting authorities and entities to design their acquisition strategy – notably in between the poles of self-production and externalisation. Canteens directly managed by the contracting authorities have a direct control on their supply line. More often than not, this leads to smaller procurements, making it easier to reach out to local suppliers. The same is true of the articulation of the procurement, with a preference however for subdivisions into lots (e.g. Article 46 of Directive 2014/24/EU). This leaves much room for pursuing further sustainable food procurement policies as shown by the experience in Dordogne (Falvo, 2023). As the *Buying Social* guide indicates, reserved procurements for sheltered workshops under Article 20 of Directive 2014/24/EU are another option (European Commission, 2021c).

One new aspect worth considering is the spotlight on resilience. Based on an earlier study, the JRC *Concepts for a sustainable EU food system* indicates that resilience requires "robust, adaptable & transformable food system", which in turn rests on elements such as diversification of farming systems, logistics resilience, diversity of the genetic pool of food plants, livestock and species in aquaculture farms, and "territorial diversity (territorial approach to food security and nutrition, short vs long chains") (Bock et al., 2022, p. 31; see also pp. 69 ff for more details). Resilience has taken a place along environmental considerations for the procurement of the technologies necessary to achieve carbon neutrality at the centre of the recent Commission proposal on the Net Zero Industry Act (European Commission, 2023). The recent shocks on the food supply chains from the pandemics and the war in Ukraine are strong reasons for resilience to be considered in food and catering procurements.

Conclusively, the Farm to Fork Strategy has triggered a process – whose exact outcomes are not yet clear – which can strongly contribute in making food and catering procurements more sustainable, while still requiring compliance with the general internal market principles of non-discrimination and transparency. Hopefully, the margins of choice of contracted authorities will not be limited by mandatory criteria not allowing more sustainable choices.

References

Andhov, M. (2021). Comment to Article 18(2). In R. Caranta & A. Sanchez Graells (Eds.), *European Public Procurement: Commentary on Directive 2014/24/EU* (pp. 199–207). Cheltenham, Edward Elgar. https://doi.org/10.4337/9781789900682

Bock, A., Bontoux, L., & Rudkin, J. (2022). *Concepts for a Sustainable EU Food System*. EUR 30894 EN. Luxembourg, Publications Office of the European Union. https://doi.org/10.2760/381319

Botta, G. (2023). Italy: Leading the Way Towards Mandatory Sustainable Public Procurement Through Minimum Environmental Criteria. In W. Janssen & R. Caranta (Eds.), *Mandatory Sustainability Requirements in EU Public Procurement Law*. Oxford, Hart (*forthcoming*).

Caranta, R. (2016). Labels as Enablers of Sustainable Public Procurement. In B. Sjåfjell & A. Wiesbrock (Eds.), *Sustainable Public Procurement Under EU Law* (pp. 99–113). Cambridge, Cambridge University Press. https://doi.org/10.1017/CBO9781316423288.006

Caranta, R. (2019). The Interplay Between EU Legislation and Effectiveness, Effective Judicial Protection and the Right to an Effective Remedy in EU Public Procurement Law. *Review of European Administrative Law*, 12(2), pp. 63–93. https://doi.org/10.7590/187479819X15840066091259

Caranta, R., & Janssen, W. (2023). Collective Reflections on the Future of Mandatory Sustainable Public Procurement. In W. Janssen & R. Caranta (Eds.), *Mandatory Sustainability Requirements in EU Public Procurement Law*. Oxford, Hart (*forthcoming*).

Dragos, D. C., & Neamtu, B. (2014). Sustainable Public Procurement in the EU: Experiences and Prospects. In F. Lichère, R. Caranta, & S. Treumer (Eds.), *Modernising Public Procurement: The New Directive* (pp. 301–335). Copenhagen, DJØF.

European Commission. (2008). *Public Procurement for a Better Environment, COM(2008) 400 Final*. https://eur-lex.europa.eu/LexUriServ/LexUriServ.do?uri=COM:2008:0400:FIN:EN:pdf

European Commission. (2016). *Buying Green! – A Handbook on Green Public Procurement*. https://ec.europa.eu/environment/gpp/pdf/Buying-Green-Handbook-3rd-Edition.pdf

European Commission. (2019a). *The European Green Deal, COM(2019) 640 Final*. https://eur-lex.europa.eu/resource.html?uri=cellar:b828d165-1c22-11ea-8c1f-01aa75ed71a1.0002.02/DOC_1&format=PDF

European Commission. (2019b). *EU Green Public Procurement Criteria for Food, Catering Services and Vending Machines. SWD(2019) 366 Final*. https://ec.europa.eu/environment/gpp/pdf/190927_EU_GPP_criteria_for_food_and_catering_services_SWD_(2019)_366_final.pdf

European Commission. (2020a). *Sustainable Europe Investment Plan. European Green Deal Investment Plan, COM(2020) 21 Final*. https://eur-lex.europa.eu/legal-content/EN/TXT/PDF/?uri=CELEX:52020DC0021

European Commission. (2020b). *A New Circular Economy Action Plan. For a Cleaner and More Competitive Europe, COM(2020) 98 Final*. https://eur-lex.europa.eu/resource.html?uri=cellar:9903b325-6388-11ea-b735-01aa75ed71a1.0017.02/DOC_1&format=PDF

European Commission. (2020c). *A Farm to Fork Strategy for a Fair, Healthy and Environmentally-Friendly Food System, COM(2020) 381 Final*.

European Commission. (2021a). *On an Action Plan for the Development of Organic Production COM/2021/141 Final/2*. https://eur-lex.europa.eu/resource.html?uri=cellar:13dc912c-a1a5-11eb-b85c-01aa75ed71a1.0003.02/DOC_1&format=PDF

European Commission. (2021b). *European Pillar of Social Rights Action Plan*. https://op.europa.eu/webpub/empl/european-pillar-of-social-rights/downloads/KE0921008ENN.pdf

European Commission. (2021c). *Buying Social – A Guide to Taking Account of Social Considerations in Public Procurement – Second edition*. 2021/C 237/01. (OJ C, C/237, 18.06.2021, p. 1, CELEX: https://eur-lex.europa.eu/legal-content/EN/TXT/?uri=CELEX:52021XC0618(01))

European Commission. (2022). *Proposal for a Directive on Corporate Sustainability Due Diligence and Amending Directive (EU) 2019/1937*. COM/2022/71 Final. https://eur-lex.europa.eu/resource.html?uri=cellar:bc4dcea4-9584-11ec-b4e4-01aa75ed71a1.0001.02/DOC_1&format=PDF

European Commission. (2023). *Proposal for a Regulation on Establishing a Framework of Measures for Strengthening Europe's Net-Zero Technology Products Manufacturing Ecosystem (Net Zero Industry Act), COM(2023) 161*. https://eur-lex.europa.eu/resource.html?uri=cellar:6448c360-c4dd-11ed-a05c-01aa75ed71a1.0001.02/DOC_1&format=PDF

European Court of Justice. (2020). *Judgment of the Court (Second Chamber) of 30 January 2020. Tim SpA – Direzione e coordinamento Vivendi SA v Consip SpA and Ministero dell'Economia e delle Finanze*. Case number = C-395/18. https://curia.europa.eu/juris/liste.jsf?num=C-395/18

European Union. (2012). Consolidated Version of the Treaty on the Functioning of the European Union. *Official Journal C*, 326, pp. 0001–0390. http://data.europa.eu/eli/treaty/tfeu_2012/oj

European Union. (2016). Consolidated Version of the Treaty on European Union. *Official Journal C*, 202, p. 13. http://data.europa.eu/eli/treaty/teu_2016/2020-03-01

European Union. (2023). *Regulation (EU) No1115/2023 of the European Parliament and of the Council of 31 May 2023 on the Making Available on the Union Market and the Export From the Union of Certain Commodities and Products Associated with Deforestation and Forest Degradation and Repealing Regulation (EU) No 995/2010*. http://data.europa.eu/eli/reg/2023/1115/oj

Falvo, C. (2023). *Report: Sustainable School Meals in Dordogne*. https://sapiensnetwork.eu/case-study-targeting-100-organic-local-and-homemade-meals-in-dordogne-secondary-schools/

FAO, Alliance of Bioversity International and CIAT and Publisher of UFRGS. (2021). *Public Food Procurement for Sustainable Food Systems and Healthy Diets* – Volume 1. Rome, FAO, Bioversity International. https://doi.org/10.4060/cb7960en. Volume 2. https://doi.org/10.4060/cb7969en. Free download.

Janssen, W. (2023). Foundations of the Paradigm Shift: Towards Mandatory Sustainability Requirements in EU Public Procurement. In W. Janssen & R. Caranta (Eds.), *Mandatory Sustainability Requirements in EU Public Procurement Law*. Oxford, Hart (*forthcoming*).

Janssen, W., & Caranta, R. (2023). *Mandatory Sustainability Requirements in EU Public Procurement Law*. Oxford, Hart (*forthcoming)*.

Lichère, F., & Sulpice, O. (2023). Mandatory Requirements in Public Procurement Law: The Role of Remedies, Courts and Public Interest Litigation. In W. Janssen & R. Caranta (Eds.), *Mandatory Sustainability Requirements in EU Public Procurement Law*. Oxford, Hart (*forthcoming*).

Marín Durán, G. (2020). Sustainable Development Chapters in EU Free Trade Agreements: Emerging Compliance Issues. *Common Market Law Review*, 57(4), pp. 1031–1068. https://doi.org/10.54648/cola2020715

Schebesta, H., & Plana Casado, M. J. (2023). Food: Mandatory EU Public Procurement Criteria for Food After the Farm to Fork Strategy. In W. Janssen & R. Caranta (Eds.), *Mandatory Sustainability Requirements in EU Public Procurement Law*. Oxford, Hart (*forthcoming*).

Schoenmaekers, S. (2023). The Influence of Sustainable Reporting Obligations on Public Purchasing. *ERA Forum*, 23, pp. 377–395.

Semple, A. (2015). *A Practical Guide to Public Procurement*. Oxford, Oxford University Press.

Semple, A. (2021). Comment to Article 67 – Award Criteria. In R. Caranta & A. Sanchez Graells (Eds.), *European Public Procurement: Commentary on Directive 2014/24/EU* (pp. 721–728). Cheltenham, Edward Elgar.

Sjåfjell, B., & Wiesbrock, A. (Eds.). (2016). *Sustainable Public Procurement Under EU Law*. Cambridge, Cambridge University Press.

Tátrai, T., & Diófási-Kovács, O. (2021). European Green Deal – The Way to Circular Public Procurement. *ERA Forum*, 22, pp. 523–539. https://doi.org/10.1007/s12027-021-00678-2.

Uysal, E. (2021). Business and Human Rights: The State as a Buyer. *European Procurement & Public Private Partnership Law Review*, 16(1), pp. 52–64. https://doi.org/10.21552/epppl/2021/1/8.

13 Transforming food systems

The role of public procurement

Marta Andhov, Sven Mikulic and Line Rise Nielsen

Introduction

The term "food system" describes the interconnected systems and processes influencing human nutrition, health, community development, agriculture and the environment. In other words, it describes everything that happens from farm to fork, paddock to plate, and soil to stomach. Food systems encompass a broad range of activities (production, aggregation, processing, distribution, consumption and disposal) and actors (producers, consumers and various intermediaries) that are inevitably embedded in and coincide with different economic, societal, physical and other environments (Nguyen, 2018; Von Braun et al., 2021). This means that food systems do not exist in a vacuum and are highly dependent on how the trade of food products is conducted, what natural resources are available to produce food, different cultures that shape dietary choices, the environment heavily affected by food production and similar factors. Still, the main objective of these complex and dynamic systems is to sustain us by keeping us fed. Unfortunately, the evidence shows that contemporary food systems have a long way to go before they achieve sustainability (Rockström et al., 2020). The global food system is the primary driver of biodiversity loss (UNEP, 2021) and a major contributor to climate change (Crippa et al., 2021). Regarding the social aspect of sustainability, specifically health, more than three billion people worldwide do not have a nutritionally adequate diet (hunger/micronutrient deficient/overweight), which is crucial for a long and healthy life (IFPRI, 2022). Simultaneously, around one-fourth of the human population works in agriculture, which is strongly related to poverty and questionable working conditions (World Bank and ILO, 2021).

Public procurement could be a strong lever for changing practices in the contemporary food system. That is due to its purchasing power through the sheer size and value of governmental contracting expenditures that amount to around 12% of the global GDP (Hafsa et al., 2021). As such, governments at all levels can create, shape and distort markets from the demand side, consequently affecting the environment and society depending on their selection from among bidders.

The European Commission's "Farm to Fork" strategy explicitly aims to make food systems fair, healthy and environmentally friendly. Green Public Procurement is envisaged as one of the mechanisms to be deployed to achieve the goals of the

DOI: 10.4324/9781003393023-16

"European Green Deal" (European Commission, 2020a). The European Commission also openly indicated its ambition to introduce minimal mandatory Green Public Procurement criteria among others in the food sector (European Commission, 2021). However, driving change and transition in food systems is a considerable challenge, and there should be no doubt that public procurement will always play a supporting role. Thus, in many cases, food-related regulation and enforcement will play a more appropriate leading role in addressing the shortcomings of current food systems.

The objective of this chapter is to illustrate how public procurement has the potential to contribute to food systems transformation. To meet this goal, the chapter is structured in the following manner. Firstly, we describe industrialised food systems and what problems arise from them. Next, we investigate the concept of agroecology as an alternative approach to food systems. Afterwards, the "interlocks" obstructing changes within the food systems are discussed, and we analyse how the EU public procurement regulatory framework, national policies and organisational capacities could challenge these interlocks. The last section provides conclusions.

Transforming food systems

Many of the challenges within the food systems can be traced to its industrialisation that started in the second half of the 19th century due to the rapidly increasing demand for food caused by the expanding world population (Smil, 2000). The industrial food systems are based on specialisation, mechanisation, high chemical and pharmaceutical input and long value chains (IPES, 2016). Such an organisation of food systems led to growing efficiency, profits, and generally cheaper food for more people than ever before (Ambikapathi et al., 2022). Simultaneously, food came to be regarded as a mere commodity, thereby disregarding its inherent non-economic, public goods characteristics, such as its pivotal role akin to clean water and air in sustaining human survival, fostering culture, preserving regional identity, and safeguarding heritage (Vivero-Pol, 2017). This shift prioritised its marketable attributes, including longevity, aesthetic appeal and the standardisation of naturally diverse food products. Consequently, food systems have for a long time been focused on producing as much food per square metre as possible at the lowest possible cost, and with a limited view of value creation (Benton & Bailey, 2019) and shifted social and ecological values and costs to the background (Huntjens, 2021).

The cost of industrial food systems is the widespread degradation of ecosystems; high GHG emissions; biodiversity losses due to the concentration on monocultures; high levels of pesticide, chemical fertilisers, and antibiotics use; and numerous other negative social and economic consequences – particularly, excessive capital intensity, farmer debt and reliance on public subsidies (IPES, 2016). These negative consequences have an immense impact on the planet and society, which is why it has been recognised that radical system-wide change is required, with "business as usual" being no longer a viable option (SAPEA, 2020). Food systems, as we know them today, must convert to more sustainable practices, i.e., *deliver food security and nutrition for all so that the economic, social and environmental bases to generate food security and nutrition for future generations are not compromised* (Nguyen, 2018).

One of the alternatives to the often-criticised industrial food systems is agroecology. This system is both a particular set of agricultural practices and a social movement (Wezel et al., 2020). The first refers to improving food systems by harnessing

Table 13.1 Agroecological Principles and Elements and their Implications for Transitioning to Sustainable Food Systems

13 Agroecological Principles
1. *Recycling*. Preferentially use local renewable resources and close as far as possible resource cycles of nutrients and biomass.
2. *Input reduction*. Reduce or eliminate dependency on purchased inputs and increase self-sufficiency.
3. *Soil health*. Secure and enhance soil health and functioning for improved plant growth, particularly by managing organic matter and enhancing soil biological activity.
4. *Animal health*. Ensure animal health and welfare.
5. *Biodiversity*. Maintain and enhance the diversity of species, functional diversity and genetic resources and thereby maintain overall agroecosystem biodiversity in time and space at field, farm and landscape scales.
6. *Synergy*. Enhance positive ecological interaction, synergy, integration and complementarity among the elements of agroecosystems (animals, crops, trees, soil and water).
7. *Economic diversification*. Diversify on-farm incomes by ensuring that small-scale farmers have greater financial independence and value-addition opportunities while enabling them to respond to demand from consumers.
8. *Co-creation of knowledge*. Enhance co-creation and horizontal sharing of knowledge including local and scientific innovation, especially through farmer-to-farmer exchange.
9. *Social values and diets*. Build food systems based on the culture, identity, tradition, social and gender equity of local communities that provide healthy, diversified, seasonally and culturally appropriate diets
10. *Fairness*. Support dignified and robust livelihoods for all actors engaged in food systems, especially small-scale food producers, based on fair trade, fair employment and fair treatment of intellectual property rights.
11. *Connectivity*. Ensure proximity and confidence between producers and consumers through promotion of fair and short distribution networks and by re-embedding food systems into local economies.
12. *Land and natural resource governance*. Strengthen institutional arrangements to improve, including the recognition and support of family farmers, smallholders and peasant food producers as sustainable managers of natural and genetic resources.
13. *Participation*. Encourage social organisation and greater participation in decision-making by food producers and consumers to support decentralised governance and local adaptive management of agricultural and food systems.

Source: Adapted from Wezel, A. et al. (2020)

natural processes such as soil diversification, creating beneficial biological interactions and synergies among the components (soil, water, human management) of agroecosystems (Gliessman, 1990) and minimising synthetic and toxic external inputs of substances such as pesticides in production (Wezel et al., 2014). The latter entails agroecology as a solution to global crises by building locally relevant and resilient food systems that strengthen the economic viability of rural areas based on short supply chains and both fair and safe food production while supporting diverse forms of smallholder food production (Wezel et al., 2020). Therefore, agroecology focuses on both the biological and social aspects of food systems by focusing on 13 principles: recycling, input reduction, soil and animal health, biodiversity, synergy, economic diversification, co-creation of knowledge, social values and diets, fairness, connectivity, land and natural resource governance and participation (Wezel et al., 2020).

Agroecology is still more of a generic framework and ideal type for evaluating specific approaches (for example, short supply chain) than a viable solution (Gowdy & Baveye, 2019). Specific approaches are envisaged to solve a defined set of problems in food systems – for example, Fairtrade primarily challenges exploitative trade practices in agriculture. Conversely, agroecology serves the holistic purpose of determining whether and how these specific approaches benefit or obstruct the sustainable development of food systems as a whole. For instance, the local approach to food systems sustainability emphasises the importance of foods produced, processed and retailed in – often subjective and poorly defined – geographical proximity to end-consumers (Kneafsey et al., 2013). Even though there might be numerous benefits to sourcing local food, that does not necessarily equate with the most sustainable food in every context (Stein & Santini, 2022). Thus, agroecological thinking probes numerous specific approaches and provides robust guidance for evaluating the sustainability of food systems across multiple dimensions. As such, it is an essential framework for making informed decisions on how to procure sustainable food.

The question arises of how to drive the transition of food systems by applying agroecological principles. As noted, *no universal pathways or win-win solutions exist as food systems transformation is highly context-specific* (Dekeyser et al., 2020). This means that transformation and prioritisation of specific approaches must be tailored to local conditions and needs – that differentiate worldwide – and there is no one-size-fits-all solution. As if complex decisions on what approaches to pick for transforming food systems are not enough, scholars also agree that certain "interlocks" keep the current industrial system in place and prevent change (De Schutter, 2017). The International Panel of Experts on Sustainable Food Systems (2016) recognised eight of these "interlocks": path-dependency, export orientation, the expectations of cheap food, compartmentalised and short-term thinking, "feed the world" narrative, measures of success and concentration of power. The instrumentalisation of public procurement mechanisms might potentially tackle some of these interlocks by reflecting agroecological principles in tender requirements and combining purchasing with broader policies and strategies.

Public procurement as leverage for sustainable food systems

Leveraging public procurement to reach policy goals is not a new phenomenon. Historically, particularly after World War II, public procurement was used as a policy tool to fulfil broader societal goals such as fighting unemployment and social inclusion and promoting innovation (Caranta, 2021). Public procurement was largely stripped from its policy tool dimension during Thatcherism and the New Public Management era. Rooted in neoliberalism that advocated for minimal state interference, public procurement became perceived simply as a bureaucratic procedure needed to fulfil a specific demand, while 'non-functional' objectives were to be set aside. In recent decades, the pendulum has been swinging back towards operationalising public procurement as a policy tool for fulfilling objectives that add value to its environment. Therefore, it is also recognised that public procurement could be leveraged to drive change in food systems (Morgan & Morley, 2014).

In many countries, food products and catering services are among the leading categories the government prioritises for including sustainability criteria in public procurement (UNEP, 2017). The reason might be the plethora of targets that can be achieved through sustainable food procurement. For instance, public buyers may demand short supply chains of sustainably grown produce and, subsequently, support the local economy in rural areas. Public kitchens can increase the provision of healthier public meals, convert to providing less meat and more seasonal vegetables, and, as such, shift food culture for future generations. In this way, sustainable food procurement sends signals about governments' ambitions for the future direction of the food system (Swensson et al., 2021), looking beyond short-term needs and considering the longer-term impacts of each purchase (Clement et al., 2016). This is possible because sustainable food procurement influences both the economic operators that are obliged to offer sustainable contract performance and the final users who gain access to sustainable food options while simultaneously having access to unsustainable food is partially restricted (Swensson et al., 2021).

The challenge in shifting towards a sustainable food system requires prioritising which sustainability aspect to tackle first, as prioritising them all is not feasible, as pursuing one (such as food security) might negatively impact another (such as soil fertility) (Dekeyser et al., 2020). However, this does not mean public buyers should continue maintaining the *status quo*.

Interlocks tackled via public procurement

As the International Panel of Food Experts (2016) recognised, eight structural dependencies or "interlocks" keep industrial food systems in place and obstruct the transition towards agroecological practices. This section highlights how employing public procurement could challenge four of these interlocks. We concentrate on "expectations of cheap food", "measures of success", "path-dependency" and "short-term thinking". These interlocks were chosen as we recognised how public procurement could challenge them. We acknowledge that the role of public procurement in challenging some of the interlocks and transforming food systems is limited. Still, as interlocks are, by definition, mutually dependent on each other, various other tools are needed in conjunction with public procurement for the complete food systems transformation to occur.

"Expectations of cheap food" and "measures of success"

Two distinct interlocks, "expectations of cheap food" and "measures of success", are grouped together for the purpose of this chapter. This is because public buyers, by altering what they value in food and how they measure the success of the food producers, inevitably shift the focus from cheap food to expecting food that does not harm its social and physical environment.

The interlock 'expectations of cheap food' refers to low prices that consumers have come to expect due to industrial mass food production. These prices often fail to account for the total costs incurred across the entire value chain, including resources utilised, fair salaries and possible negative externalities occurring (IPES, 2016). Incorporating these costs in pricing can result in higher direct food prices but a lower total societal cost. As with individual consumers, the same logic applies to public buyers who must be aware of these additional costs that, if not accounted for, inevitably manifest elsewhere. For instance, the costs of not having nutritious and healthy diets may manifest in higher healthcare expenditures borne by the public pocket in the future.

The interlock "measures of success" points out that the current metrics by which food producers are recognised as successful are indispensable from industrialised food systems logic. Food producers are considered successful based on the high total yields of specific crops, productivity per worker, total factor productivity, and similar metrics (IPES, 2016). Instead, new indicators of success must go beyond narrowly understood cost-benefit analyses and account for and measure food products' impacts on social and physical environments.

Public buyers can influence how the food and producers are valued by incorporating sustainability requirements in different stages of public procurement. In what follows it is discussed how public buyers can change what they value in food via technical specifications, award criteria and contract performance conditions. By reflecting on what public buyers value in different stages of public procurement, food producers are signalled about measures of success required by agroecology rather than industrial agriculture.

EU public procurement law makes clear that technical specifications of products or services to be procured may include sustainability considerations (Directive 2014/24/EU, article 74). One way of drawing up technical specifications for sustainable food procurement is by prohibiting certain ingredients or requiring reduction of them in the purchased food (for example, artificial additives, trans fats, palm oil). Another way can be by including sustainability requirements such as fresh, seasonal and made-from-scratch food. Labels and standards indicating a level of sustainability and quality of food products can also be utilised. For instance, organic certification is a regulated and well-established tool for sustainable food procurement as it provides an objective and verifiable measurement of sustainability that encapsulates elements of agroecology while also *a priori* providing bidders with clear rules and sufficient clarity of expectations. It is recognised that organic certifications are used as a measure of environmentally sound products (Molin et al., 2021). By drawing technical specifications in such a way, minimum levels of environmental performance are set for all bidders (Schebesta, 2016). The market is signalled about changing measures of success for economic operators, i.e., farmers are perceived as successful if they produce organically. This is needed as many current policies are leaning heavily towards promoting organic supply through the provision of farm subsidies rather than stimulating demand for organic food products (Daugbjerg, 2023).

The usage of labels lessens the work for public buyers – in terms of introducing comprehensive requirements for bidders without repeatedly specifying every single detail that must be satisfied. However, it potentially limits access for small and medium enterprises (European Commission, 2020b) and may result in higher costs of products procured (Sönnichsen & Clement, 2020). In other words, each certification requires economic operators to incur certain costs necessary for obtaining and maintaining a label. Bearing these additional costs can be unattainable for some economic operators – specifically small and medium enterprises – which prevents their access to public procurement markets and may hinder sustainable food procurement. Moreover, the economic operators who offer sustainably certified food products account for this cost in their production, which results in price premiums associated with such purchases. This, in turn, deters public buyers from introducing such requirements as they operate with only limited budgets. Therefore, only gradual inclusion of such requirements is feasible as public buyers must be able to find budgetary space to pay price premiums for sustainable food products.

Award criteria, which provide the metric to compare submitted bids, have the potential to tackle both expectations of cheap food and measures of success for farmers. The most economically advantageous tender (MEAT) criteria envisage the evaluation of bids that goes beyond the lowest-cost approach that perpetuates the expectation of cheap food. When it comes to food and catering services, these criteria can include quality aspects (such as diversification of protein sources, fresh, seasonal, high nutritional value), technical merit (such as experience in the provision of healthy and sustainable catering), environmental (such as food waste donation agreements, climate-friendly delivery conditions), and social characteristics (such as Fairtrade). As such, the most economically advantageous approach shifts the focus from simplistic price evaluations to other parameters that public buyers value regarding food.

Finally, in contract performance conditions, public buyers may reiterate criteria expressed in technical specifications or award criteria (Boyano et al., 2019). They add an additional layer of enforcement, which can be attributed to penalty clauses for non-compliance (Andhov, 2021a). In this way, public buyers ensure that sustainability requirements, as reflected in technical specifications and award criteria, are respected by economic operators in the contract performance phase.

Changing "measures of success" for economic operators and "expectations of cheap food" for public buyers goes beyond drawing up tender requirements contract clauses. Integrating sustainable food procurement into broader strategies amplifies its potential in promoting sustainable food systems. For example, when public school kitchens collaborate with students to prepare school meals, it can serve an educational role and be a tangible illustration of the workings and significance of a sustainable food system. In this way, students can gain insights into regionally sourced produce's historical, cultural, and heritage aspects, understand its seasonal availability and appreciate its health benefits. Such educational purpose has been a key argument for Copenhagen Municipality in procuring specific crops such as apples, potatoes, asparagus and different berries. Similarly, in their tenders, the Copenhagen Hospitality School focuses on educating future generations of kitchen staff about the connection between urban and rural areas and minimising food waste, for example, by using all cuts of meat. Such tender documents required economic operators to reflect in their bids on how they would contribute to the education of the students by sharing their expertise, for example, through arranging visits and dialogue with producers and end-consumers. Thus, bidders are perceived as successful based not only on the food products they are able to supply but also on the added value in terms of educating students about their sustainable practices.

"Path-dependency" and "short-term thinking"

The interlock "path-dependency" describes the creation of policies that lock the system into specific technologies and power relationships (for example, industrial agriculture) instead of leaving open future possibilities (for example, preserving or enhancing soil quality) (Gowdy & Baveye, 2019). Regulations function in the same manner by either allowing and/or incentivising certain paths or *vice versa*. The interlock "short-term thinking" refers to a trend in policymaking that emphasises policies with immediate results and benefits. It is usually a result of politicians opting out of policies whose rewards will not materialise within the same electoral cycle.

Public procurement policymaking in the EU happens on at least four levels of governance: EU, national, municipal and organisational. At the EU level, sustainable food procurement is governed by public procurement directives and sustainable procurement policies such as "EU Green Public Procurement criteria" for food or catering services or food-specific policies (see Section 2.4, "Farm to Fork" strategy). Member States govern via national public procurement acts, transposing the EU public procurement directives and their own policies. Municipalities govern via their own food and procurement strategies, and individual public procurers via their discretionary power in designing tenders and choosing what to buy. All public procurement governance instruments at different levels can either encourage the current focus on industrialised food systems or promote a different approach – aimed at promoting long-term sustainability. Correspondingly, if public organisations are focused on short-term thinking, this will be reflected in their public procurement policies.

The EU aims to break the unsustainable path of food systems, which can be seen in its public procurement policies and rules. Explicit permission for conducting sustainable food procurement was already found in Directive 2004/18/EC (for example,

Articles 23 and 53 and Recitals 33 and 46), and it has been reconfirmed in Directive 2014/24/EU. In 2008 food and catering services were identified as priority sectors for the green transition that should be tackled via public procurement (European Commission, 2008). As it was recognised that there was a lack of legal and organisational guidance for public buyers on how to procure sustainably, the EU Green Public Procurement criteria were created. The latest edition of Green Public Procurement criteria for food and catering services was published in 2019 and aimed at providing clear and scientifically backed criteria that would lead to more sustainable food procurement (Boyano et al., 2019). It is evident that elements of agroecology, both as a practice and as a social movement, are reflected in EU Green Public Procurement criteria. It can be concluded that at the EU governance level, the actors are proactive in terms of facilitating sustainable food procurement. This can be clearly seen in the EU communications, laws and strategies. Thus, the EU challenges "path-dependency" in food systems and displays 'long-sightedness' in envisaging future food systems.

However, the actual outcomes of the EU's "long-sightedness" depend on implementing the same policies within Member State governance and later at the local level. Thus, aside from navigating complex legal public procurement frameworks, utilising public procurement for transforming food systems requires holistic long-term investments and long-sighted political leadership at national levels. The latter has been practised in Denmark, where sustainable food procurement was part of a broader, politically-backed, holistic action plan for stimulating the organic conversion of agriculture by creating a higher demand for organic food.

In 2011, the national goal of 60% organic public kitchens was introduced, resulting in a significant increase in organic public procurement among public kitchens (Sørensen et al., 2016). IFOAM (2020) recognised that the reasons for Denmark's success were the setting of clear political goals, financing of education in public kitchens, the establishment of national labels for organic cuisine[1] and financial support from the public Fund for Organic Agriculture for capacity building in the organic sector NGOs (especially the Copenhagen House of Food). More precisely, Denmark invested in educating kitchen staff on sustainable practices rather than increasing the budget to pay a premium for organic food products. These included reducing food waste and meat usage, seasonal menus and cooking from scratch (Mikkelsen & Sylvest, 2012; Sørensen et al., 2016). On a more granular level, the city of Copenhagen rolled out a holistic, organic strategy that has been supported by a unanimous city council for more than 12 years with a goal for all public kitchens in the municipality to be at least 90% organic (City of Copenhagen Food and Meal Strategy, 2019). Such national and local policies challenge the path of dependence on conventional agriculture by considerably stimulating demand for organic food via public procurement.

Lastly, beyond the EU, national and local policies, the uptake of sustainable food procurement still largely depends on the discretionary decisions of public buyers on what and how to buy (Andhov, 2021b). For instance, public buyers often advertise large and bundled contracts to enhance efficiency and reduce administrative burden. In the sustainable food procurement context, aggregated contracts limit Small and Medium Enterprise's participation as they lack the capacity to execute them (Miranda, 2018). Even though it cannot be *a priori* asserted that the small-farm holders follow agroecological principles, they do have predispositions that make them susceptible to them (Ebel, 2020). Therefore, actions such as dividing contracts into lots are needed to challenge the interlock of path-dependency on industrial food systems and provide small producers access to the public procurement market.

Conclusion

The current organisation of food systems is not sustainable, and the employment of agroecological practices can change this trajectory. As a policy and regulatory tool, public procurement plays a supporting role in transforming food systems by creating considerable demand for sustainable food. However, it is evident that many approaches to conducting sustainable food procurement exist and that some structural dependencies (interlocks) in food systems prevent change. Thus, when choosing which approach to take, public buyers shall familiarise themselves with agroecological principles to make an informed decision about what is and is not a sustainable food requirement in each specific context. In addition to this, public buyers have tools at their disposal to tackle interlocks that prevent change. By incorporating various sustainability metrics in different stages of public procurement, public buyers can signal a change in measures of success for bidders. Further, by employing the most economically advantageous tender criteria instead of the lowest-price evaluation method, public buyers shift focus from expecting cheap food to expecting food that adds value to its environment. Still, it is also recognised that sustainable food procurement has to exist as a part of holistic and strongly politically-backed long-term strategies at all governance levels to be successful in challenging the reliance on industrial food systems.

Note

1 Thirty to sixty percent usage of organic raw ingredients and beverages in kitchens for bronze label, 60–90% for silver and 90–100% organic share for gold. To learn more about Danish "Organic Cuisine Label" see www.oekologisk-spisemaerke.dk/horeca-en.

Bibliography

Alexander, P., Brown, C., Arneth, A., Finnigan, J., Moran, D., & Rounsevell, M. D. A. (2017). Losses, inefficiencies and waste in the global food system. *Agricultural Systems*, *153*, 190–200. doi: 10.1016/j.agsy.2017.01.014

Ambikapathi, R., Schneider, K. R., Davis, B., Herrero, M., Winters, P., & Fanzo, J. C. (2022). Global food systems transitions have enabled affordable diets but had less favourable outcomes for nutrition, environmental health, inclusion and equity. *Nature Food*, *3*(9), 764–779. doi: 10.1038/s43016-022-00588-7

Andhov, M. (2021a). Article 70 contract performance conditions. In *European Public Procurement* (pp. 747–754). Edward Elgar Publishing. doi: 10.4337/9781789900682.00079

Andhov, M. (2021b). *Contracting Authorities and Strategic Goals of Public Procurement – a Relationship Defined by Discretion?* doi: 10.5040/9781509919512.ch-006

Benton, T. G., & Bailey, R. (2019). The paradox of productivity: Agricultural productivity promotes food system inefficiency. *Global Sustainability, 2*, e6. doi: 10.1017/sus.2019.3

Boyano, L. A., Espinosa Martinez, M. N., Rodriguez Quintero, R., Neto, B., Gama Caldas, M., & Wolf, O. (2019). *EU GPP Criteria for Food Procurement, Catering Services and Vending Machines*. JRC. Retrieved from https://data.europa.eu/doi/10.2760/748165

Caranta, R. (2021). *Public Procurement for the SDGs – Rethinking the Basics*. doi: 10.2139/ssrn.4038840

City of Copenhagen. (2019). *The City of Copenhagen's Food Strategy*. The City of Copenhagen. Retrieved from https://maaltider.kk.dk/sites/default/files/2022-06/The%20City%20of%20Copenhagen%20Food%20Strategy%202019.pdf

Clement, S., Watt, J., & Semple, A. (2016). *The Procura+ Manual, A Guide to Implementing Sustainable Procurement*. ICLEI – Local Governments for Sustainability, European Secretariat.

Crippa, M., Solazzo, E., Guizzardi, D., Monforti-Ferrario, F., Tubiello, F. N., & Leip, A. J. N. F. (2021). Food systems are responsible for a third of global anthropogenic GHG emissions. *Nature Food, 2*(3), 198–209. doi: 10.1038/s43016-021-00225-9

Daugbjerg, C. (2023). Using public procurement of organic food to promote pesticide-free farming: A comparison of governance modes in Denmark and Sweden. *Environmental Science & Policy, 140*, 271–278. doi: 10.1016/j.envsci.2022.12.012

Dekeyser, K., Rampa, F., D'Alessandro, C., & Bizzotto Molina, P. (2020). *The Food Systems Approach in Practice: Our Guide for Sustainable Transformation*. European Centre for Development Policy Management.

De Schutter, O. (2017). The political economy of food systems reform. *European Review of Agricultural Economics, 44*(4), 705–731. doi: 10.1093/erae/jbx009

Directive 2004/18/EC. On the coordination of procedures for the award of public works contracts, public supply contracts and public service contracts. *European Parliament and Council*. Retrieved from http://data.europa.eu/eli/dir/2004/18/oj

Directive 2014/24/EU. On public procurement. *European Parliament and Council*. Retrieved from http://data.europa.eu/eli/dir/2014/24/oj

Ebel, R. (2020). Are small farms sustainable by nature? – Review of an ongoing misunderstanding in agroecology. *Challenges in Sustainability, 8*(1), 17–29. doi: 10.12924/cis2020.08010017

European Commission. (2008). *Communication from the Commission to the European Parliament, the Council, the European Economic and Social Committee and the Committee of the Regions: Public Procurement for a Better Environment*. Retrieved from https://eur-lex.europa.eu/legal-content/EN/TXT/?uri=celex%3A52008DC0400

European Commission. (2020a). *Communication From the Commission to the European Parliament, the Council, the European Economic and Social Committee and the Committee of The Regions: A Farm to Fork Strategy for a Fair, Healthy and Environmentally-Friendly Food System*. Retrieved from https://eur-lex.europa.eu/legal-content/EN/TXT/?uri=celex%3A52020DC0381

European Commission. (2020b). Directorate-general for internal market, industry, entrepreneurship and SMEs. In Bas, P., Hausemer, P., & Kruger, T. (Eds.). *Analysis of the SMEs' Participation in Public Procurement and the Measures to Support it: Final Report*. Publications Office. Retrieved from https://data.europa.eu/doi/10.2873/417621

European Commission. (2021). *Communication From the Commission to the European Parliament, the Council, the European Economic and Social Committee and the Committee of the Regions on an Action Plan for the Development of Organic Production*.

Retrieved from https://eur-lex.europa.eu/legal-content/EN/TXT/?uri=CELEX%3A52021DC0141R%2801%29

FAO Agrifood Economics Division (Ed.). (2022). *The State of Food Security and Nutrition in the World 2022*. FAO. doi: 10.4060/cc0639en

Gliessman, S. R. (1990). Agroecology: Researching the ecological basis for sustainable agriculture. In: Gliessman, S. R. (Eds.), *Agroecology. Ecological Studies* (Vol. 78). Springer. doi: 10.1007/978-1-4612-3252-0_1

Gowdy, J., & Baveye, P. (2019). An evolutionary perspective on industrial and sustainable agriculture. In *Agroecosystem Diversity – Reconciling Contemporary Agriculture and Environmental Quality* (pp. 425–433). Academic Press. doi: 10.1016/B978–0–12–811050–8.00027–3

Hafsa, F., Darnall, N., & Bretschneider, S. (2021). Estimating the true size of public procurement to assess sustainability impact. *Sustainability*, *13*(3), 1448. doi: 10.3390/su13031448

Huntjens, P. (2021). *Towards a Natural Social Contract: Transformative Social-Ecological Innovation for a Sustainable, Healthy and Just Society* (p. 205). Springer Nature.

IFOAM. (2020). *Best Practice in Organic Public Procurement: The Case of Denmark A Successful Business Case for Organic Public Procurement. With useful lessons for other nations. International Federation of Organic Agriculture Movements – Organics International.* Retrieved from www.organicseurope.bio/content/uploads/2021/06/IFOAMOE_Best-Practice-in-Organic-Public-Procurement_The-case-of-Denmark.pdf?dd

IFPRI. (2022). *2022 Global Food Policy Report: Transforming Food Systems After COVID-19*. International Food Policy Research Institute. doi: 10.2499/9780896293991

IPES. (2016). *From Uniformity to Diversity: A Paradigm Shift From Industrial Agriculture to Diversified Agroecological Systems*. Louvain-la-Neuve. Retrieved from www.ipes-food.org/_img/upload/files/UniformityToDiversity_FULL.pdf

Kneafsey, M., Venn, L., Schmutz, U., Balázs, B., Trenchard, L., Eyden-Wood, T., Bos, E., Sutton, G., & Blackett, M. (2013). *Short Food Supply Chains and Local Food Systems in the EU (EUR 25911 EN; JRC Scientific and Policy Reports)*. Publications Office of the European Union. doi: 10.2791/88784

Kugelberg, S., Bartolini, F., Kanter, D. R., Milford, A. B., Pira, K., Sanz-Cobena, A., & Leip, A. (2021). Implications of a food system approach for policy agenda-setting design. *Global Food Security*, *28*, 100451. doi: 10.1016/j.gfs.2020.100451

Mikkelsen, B. E., & Sylvest, J. (2012). Organic foods on the public plate: Technical challenge or organisational change? *Journal of Foodservice Business Research*, *15*(1), 64–83. doi: 10.1080/15378020.2011.650541

Miranda, A. (2018). *Public Food Procurement From Smallholder Farmers: Literature Review and Best Practices (ISSN: 1812–108x)*. Food and Agriculture Organization of the United Nations and the United Nations Development Programme.

Molin, E., Martin, M., & Björklund, A. (2021). Addressing sustainability within public procurement of food: A systematic literature review. *Sustainability*, *13*(23), 13395. doi: 10.3390/su132313395

Morgan, K., & Morley, A. (2014). The public plate: Harnessing the power of purchase. In *Sustainable Food Systems: Building a New Paradigm*. Routledge.

Nguyen, H. (2018). *Sustainable Food Systems: Concept and Framework. Brief Based on a Training Course "Introduction to Sustainable Food Systems and Value Chains"*. FAO.

Rockström, J., Edenhofer, O., Gaertner, J., & DeClerck, F. (2020). Planet-proofing the global food system. *Nature Food*, *1*(1), 3–5. doi: 10.1038/s43016–019–0010–4

SAPEA. (2020). *A Sustainable Food System for the European Union*. Science Advice for Policy by European Academies (SAPEA). doi: 10.26356/sustainablefood

Schebesta, H. (2018). Revision of the EU green public procurement criteria for food procurement and catering services – certification schemes as the main determinant for public sustainable food purchases? *European Journal of Risk Regulation*, *9*(2), 316–328. doi: 10.1017/err.2018.24

Smil, V. (2000). *Feeding the World: A Challenge for the Twenty-First Century*. MIT Press. doi: 10.7551/mitpress/2977.001.0001

Sönnichsen, S. D., & Clement, J. (2020). Review of green and sustainable public procurement: Towards circular public procurement. *Journal of Cleaner Production*, *245*, 118901. doi: 10.1016/j.jclepro.2019.118901

Sørensen, N. N., Tetens, I., Lassen, A. D., & Løje, H. (2016). *Organic Food Conversion in Danish Public Kitchens: The Effects of the Danish Organic Action Plan 2020 on Organic Public Procurement and Wellbeing at Work*. National Food Institute, Technical University of Denmark. Published PhD thesis. Retrieved from http://orbit.dtu.dk/files/124231485/PhD_Thesis_Nina_N_rgaard_S_rensen_print_final.pdf

Sørensen, N. N., Tetens, I., Løje, H., & Lassen, A. D. (2016). The effectiveness of the Danish organic action plan 2020 to increase the level of organic public procurement in Danish public kitchens. *Public Health Nutrition*, *19*(18), 3428–3435. doi: 10.1017/S1368980016001737

Stein, A. J., & Santini, F. (2022). The sustainability of "local" food: A review for policy-makers. *Review of Agricultural, Food and Environmental Studies*, *103*(1), 77–89. doi: 10.1007/s41130-021-00148-w

Swensson, L. F. J. S., Hunter, D. H., Schneider, S. S., & Tartanac, F. T. (2021). Public food procurement as a game changer for food system transformation. In *Public Food Procurement for Sustainable Food Systems and Healthy Diets* (1st ed., Vol. 1, pp. 1–24). FAO, Biodiversity International.

UNEP. (2017). *Global Review of Sustainable Public Procurement*. United Nations Environment Program.

UNEP. (2021). *Our Global Food System is the Primary Driver of Biodiversity Loss*. UN Environment. Retrieved December 10, 2022 from www.unep.org/news-and-stories/press-release/our-global-food-system-primary-driver-biodiversity-loss

Vivero-Pol, J. (2017). Food as commons or commodity? exploring the links between normative valuations and agency in food transition. *Sustainability*, *9*(3), 442. doi: 10.3390/su9030442

Von Braun, J., Afsana, K., Fresco, L., Hassan, M., & Torero, M. (2021). Food systems – definition, concept and application for the UN food systems summit. *Science and Innovation*, *27*. Retrieved from www.un.org/sites/un2.un.org/files/2020/12/food_systems_paper-draft_oct-25.pdf

Wezel, A., Casagrande, M., Celette, F., Vian, J.-F., Ferrer, A., & Peigné, J. (2014). Agroecological practices for sustainable agriculture. A review. *Agronomy for Sustainable Development*, *34*(1), 1–20. doi: 10.1007/s13593-013-0180-7

Wezel, A., Herren, B. G., Kerr, R. B., Barrios, E., Gonçalves, A. L. R., & Sinclair, F. (2020). Agroecological principles and elements and their implications for transitioning to sustainable food systems. A review. *Agronomy for Sustainable Development*, *40*, 1–13. doi: 10.1007/s13593-020-00646-z

World Bank & ILO. (2021). *Employment in Agriculture (% of Total Employment)* [Dataset]. The World Bank. Retrieved from https://data.worldbank.org/indicator/SL.AGR.EMPL.ZS

14 A paradigm shift in public food service

Maurizio Mariani

Introduction

Why do we need to address a paradigm shift in our food system and in the public procurement of food and catering services, whether for schools, universities, hospitals or nursing homes?

Our food system has been highly developed and focused heavily on food safety, which remains a central issue, but it has become too industrialised. Too many ultra-processed foods are present in the food habits of Europeans. We already know the consequences. Unfortunately, all of this has developed to the detriment of the environment and our health and very often to the detriment of human labour. It developed at a time in history when the word resilience was still only relegated to mechanics.

This chapter also describes case studies such as that of the Dordogne Department (France) which has revolutionised its school meals system, starting by placing the cook back at the centre and enhancing his social role; we want to expand our horizons.

Move away from legal technicalities for a moment, and focus on how, even upstream of a tender, we should be thinking and investing in order to create the prerequisites that will later become the prerogatives of the new food system and a new European strategy on sustainable public procurement of catering services.

At the end of this chapter, we highlight the symbiotic relationship between food education and food literacy that are the linchpin for cultivating a society characterised by health-conscious, informed and responsible consumers.

Together, they unravel the complexities of modern food systems, promote healthy eating habits and contribute to the mitigation of diet-related health issues such as obesity, diabetes and cancer.

Facing a dilemma

Our food system has been experiencing a profound transformation with enormous consequences on people's and the planet's health, starting with the end of World War II. The change in how we produce and consume food has insidiously affected many areas of life – climate change, environment and biodiversity, nutritional quality and human health, labour, water and land use. This is the dilemma we are going to face in the coming years. (Pollan, 2007).

DOI: 10.4324/9781003393023-17

Health experts warn that the increasing popularity of ultra-processed food will lead to adverse effects such as obesity and poor health. Research by global nutrition experts and national health councils reveals that the scale of our food evolution, from farm-fresh to highly processed, will directly affect health. (Kliemann et al., 2023). "Real food" has been replaced by salty snacks, sugary cereals, industrially-made bread and desserts, ready meals and reconstituted meats alongside sweetened soft drinks.

In Europe, the food system's transformation process should be put in the context of two demographic issues: an aging population and a "food habits" evolution due to ethical and environmental issues, especially in the youth generation. These are long-term trends that began several decades ago in Europe.

The health problems and nutritional needs of an aging population differ from those of younger demographics, and the transformation of our food system has to account for that.

It is certainly possible to combine food and the environment without running the risk of dividing today's society, especially within schools and universities, and that only by promoting proper food and environmental education and investing in sustainable practices in order to create a more equitable, healthy and sustainable food system for all.

The public food services

In Europe, about 90 million meals a day are consumed in canteens (public and private served meals combined), and more than 50% of which are in the public food service sector "PFS" (schools, hospitals, nursing homes. . .). This market segment represents an excellent opportunity for local administrations to drive toward sustainability in the entire food service market and foster the new concept of a regional food system.

This chapter will investigate how and where food is procured, the nature of food that is sourced, how it is prepared and consumed, and how it can be used to produce multiple benefits for society, the economy and the environment, which are the three main pillars of Sustainable Development.

Supporting local and small-scale producers can prioritise sourcing from local and small-scale producers, which helps to strengthen the local economy and reduce the environmental impact of transporting food over long distances, setting standards for food production that prioritise animal welfare, worker rights and environmental sustainability.

Public food service procurement can also be used to educate and engage the public about sustainable and healthy food choices. This can include offering educational programmes, promoting food literacy and encouraging public participation in food-related decision-making.

Procurement policies aimed at rewarding the sustainability and healthy qualities of raw ingredients used in the preparation of the meals are also a lever to fight climate change. The knowledge exchange and networking among professionals become a significant lever to pursue sustainability and foster innovation, socially even more than technologically. Procurement should consider the role of the kitchen brigade as a whole, starting from the cooks to the caretaker; all of them perform an essential job, which must re-take their rightful social role and give them a new motivation.

To provide good meals more is needed than to purchase good ingredients; we need the knowledge of the cook and the skill of the entire brigade, as a whole. We do not have to "assemble" a recipe; we have to perform a recipe to celebrate each ingredient and serve a delicious dish that tells a story and gives an experience to those who eat it.

Public food service in Europe is currently split into two large worlds. On the one hand, there are the local authorities who have had the vision to maintain the service "in-house", i.e., managing it directly; on the other hand, there are those who, for economic reasons or strategic choices, have preferred to contract it out to external companies.

There are also different operating models, with centralised kitchens for three- to four thousand meals a day and veritable "factories" producing, instead "assembling" up to 50 thousand meals a day, whether "fresh-hot" or "fresh-cold" delivery, using in large part processed and ultra-processed food.

In the name of the so-called "economies of scale", the size of external kitchens has been growing, often working for several clients and certainly more competitive than small kitchens, but at the expense of artisanal production and very often of quality.

In this context, there are still schools with in-house kitchens that work with local products, generating significant economic impact on the local area and guaranteeing the artisanal production and freshness of raw ingredients that large kitchens cannot ensure.

The Dordogne Department case study: a dream come true

The innovative management model is based on a systemic approach that has made it possible to connect the real "nutritional" needs of school canteens with local food production, re-centring activities on the enhancement of the skills of each actor in the supply chain and all stakeholders, putting human labour back at the centre of the local food system.

In the kitchens in particular, taking example from what was professed by the famous chef Auguste Escoffier, with a few weeks of training in "immersion", redefining precisely the role of each staff member and motivating the whole team, thanks also to the purchase of new equipment they have gone from producing meals starting from semi-finished industrial products to cooking with local, fresh and seasonal raw ingredients (food items).

All this is accompanied by several educational moments to support students in transitioning from eating industrial products to a local, healthy and nutritious diet, which offers them delicious meals daily.

In the various specifications for procurement, they have moved from rigidly purchasing "products" to buying food items, pre-defining the various food categories and their annual volume, and setting a fair price that remunerates smallholder producers. Only at the moment of weekly purchase is it defined which exact food item will be purchased within the defined food category; this flexibility is one of the central pillars of the project, capable of adapting to the territory offer and increasing its resilience.

Dordogne is an exemplary case; thanks to the political vision of its President Germinal Peiro and his motivated and skilled staff, it has succeeded in making school canteens a pedagogical moment and an example to follow, not only for having launched an ambitious programme of reorganisation school canteens but for having created around them a value for the territory; a commercial outlet for smallholders local producers who are engaged and have a key role no longer as mere spectators within the food system of the territory, through a genuine win-win relationship.

An example of this is the college of Nontron, where before this "delicious revolution", as Olivier Roellinger defined it, only four teachers ate in the canteen, and today 70 of them (all) eat in the canteen with the children and have become the first testimonials, guarantors of the project itself.

The department has been able to build a model of a healthy and local diet, mainly based on local organic agriculture, building a real food community around the project itself that guarantees its accomplishment.

The department has created the conditions on which the individual schools, which in some aspects are independent of it, approve the project and create the food community that supports its implementation.

This model of school canteen management not only offers healthy and local food, strengthening the economy of the territory, but it also contributes concretely to the reduction of climate-changing emissions related to the food system, reducing the kilometres travelled by food, significantly reducing the amount of packaging needed for their transport and thanks to the implementation of a composting system to return raw material to agriculture, creating an entirely circular economy system.

Finally, all this without spending a cent more, without eliminating meat from the menu while rediscovering legumes and vegetables, in part forgotten, and using herbs and spices instead of salt to enrich the taste of the recipes.

A truly "delicious revolution" which, considering its scope, will have an impact on many families living within the department, at the very least, given the 14,000 students (only from the colleges), the staff involved and the family coefficient of 3.66 people (per family with children) it will reach over 15% of the population living in the department.

A story of will

The Dordogne Department voted for a "Eat Local in Public Foodservice" roadmap in June 2016. This integrated approach ranges from production support to the use of products, including marketing and distribution.

The Dordogne is a poor department, among the ten poorest in France in terms of median income. As part of its compulsory responsibilities, the Dordogne

Department directly manages the school meals of 35 colleges. These 35 schools represent 14,077 students, of which nearly one-third are on scholarship. In 2019, 1.9 million meals were served in these establishments.

The department council has chosen to keep a kitchen in each establishment and to have its agents prepare all meals in-house, with the support of two cook trainers and a nutritionist. They use the National Agrilocal ordering platform to facilitate exchanges between colleges and suppliers in compliance with public procurement rules.

They also contribute to creating the "Manger Bio Périgord", a logistic platform gathering local organic producers to facilitate market access to small-scale producers.

The farming system

The Dordogne has the chance to have a farming system that offers a wide range of quality fruits, vegetables, cereals and meats. Some of the producers transform their products on the farm. The food artisans and the food processing companies work in this large local production and participate, with farmers, in the gastronomic reputation of the Périgord, which millions of tourists discover with delight every year.

The useful agricultural surface of the Dordogne (SAU) is 353.100 hectares, of which 42,372 hectares. (12%) are organic. From a calculation made regarding the nutritional needs of school children, it was found that the cultivation needs to satisfy a 100% organic and local supply of all school meals is around 169 hectares, of which 41 hectares are for the colleges alone.

The vision

The departmental political majority is committed to a 100% organic and local food supply for the colleges for which the department is responsible.

This commitment aims to:

- Protect human health
- Relocate the economy by having a holistic approach to the food issue
- Cook on a realistic scale: Respect the operational thresholds beyond which it becomes impossible to cook artisanally
- Respect the seasonality of products, the environment and animal welfare
- Control costs
- Fair prices for the producers
- Promote food education: Children eating 100% organic and local meals are the consumers/citizens of tomorrow

To reach this objective, the department has implemented a multifaceted approach similar to a quasi-integrated approach from field to plate and beyond.

Eleven establishments have been supported and offer 100% organic and local meals. These schools have been certified by ECOCERT and are the first schools in France to obtain this label and the level of 100% organic and local products.

In concrete terms, the following actions have been carried out in the last four years (non-exhaustive list):

- An exhaustive inventory of the offer and demand of organic products within the territory of the department
- Support for the development of local organic production
- Matching of schools' canteen needs with farmers' production
- Work on drafting public procurement contracts and, more generally, on the public procurement policies of the schools
- Supporting an association of organic producers in implementing and managing a marketing/distribution platform of organic products for school canteens
- Development of training in 100% organic and local cuisine by a chef trainer for college chefs
- Accompaniment of the chefs on the food plan, dietetics and the menus design
- Support for schools in the integrated implementation of 100% organic and local experiments in schools with the entire educational community (internal and external communication)
- Upgrading of kitchen equipment adapted to the preparation of organic and local products
- Waste reduction and recycling/composting

The resources mobilised are mainly the department's resources (agriculture service, education department, environment and sustainable development department) which have been strengthened by the recruitment of three people dedicated to this objective: two chef trainers specialised in 100% organic and local food and a dietician.

To further support, the department has chosen the "Collectif Les pieds dans le plat", with which it recently founded the cooperative of collective interest "Nourrir l'avenir" to disseminate nationwide the experience carried out.

Enhance and re-localise the farming system

The department's approach is based on stimulating the demand for organic and local products through school canteens. However, this demand must be met according to the two main criteria of local origin and the organic nature of the products.

The department contributes to the upstream organisation of agricultural production and supply chains to achieve this objective. This is done in particular by:

- Identification of production volumes by product type (food items) and territory
- Support for equipment investment for developing farms
- Advice and support for farm management and set-up
- Support for the development of a logistics platform for the marketing and consolidation of organic and local products
- The co-financing of tools for processing local products (vegetable gardens, slaughterhouses/cutting room)

All these measures have the effect of improving the visibility of the products produced in the region, encouraging production through public procurement, fair remuneration for farmers, limiting the carbon impact of production, encouraging the use of ultra-short circuits, combating a form of agricultural decline in rural areas, relocating production while respecting seasonality, diversifying outlets for farmers and offering a medium/long-term vision through long-term procurement contracts.

There is a story of friendship and trustworthy relationship beyond this challenging project; the president of the department met a young chef working in a small village of the department in a canteen school. Passion for rugby was the common space between them. After another person joined the two, a dietician, and with this "triple skill" of one visionary politician, an empowered chef and a clever dietician, the dream came true, after less than two years, thanks to the engagement of the whole department staff.

Lesson learned or still a dream?

Obviously, for reasons of geography, space and affordability of the service, this model can be adapted to large cities only by creating what we call "neighbourhood kitchens" or "Kitchen Food Hubs".

The Kitchen Food Hub "KFH" is a place where raw ingredients can be aggregated, processed into delicious meals and shared with consumers and institutions. It becomes a key component to the success of the urban food system, by reaching out and serving meals to the community and educating people about the benefits of real, good food. Furthermore, it will connect Food Dots and minds in a win-win relationship able to create delicious meals, to foster a healthy & sustainable regional food system.

The Kitchen Food Hub is not just part of the food supply chain but also part of the local "values network". So while providing infrastructure to local farmers and businesspeople is a vital part of what the Kitchen Food Hub does, an equally important role is to engage people who feel isolated or lack skills in nutrition and cooking. Facilitating a wide-ranging combination of life and employment skills, personal support, healthy eating and togetherness is what turns a commercial kitchen into a hub for socialising, the pleasure of eating healthy food.

Around many cities, and in some cases inside the city, there are a lot of small food producers (farmers, fishermen, artisanal food processors, etc.) that are largely excluded from the urban market. Being small-scale makes it impossible for them to compete with larger food service companies. All these entities are what we call Food Dots, which need to be connected with other dots that consume food, namely school canteens, hospitals, nursing homes, prisons, or individual public/private employees holding a meal voucher.

The Kitchen Food Hub connects the Food Dots through distinct but interacting physical, digital, and logistical sub-systems based on the following components:

- Production unit (kitchen)
- Aggregation, processing and distribution centre

- Kitchen incubator
- Canteen
- Farmers market and meeting room

The production kitchen transforms raw ingredients into meals. The Kitchen Food Hub intends to revalue the meals provided to urban communities by public/private kitchens, traditionally produced and delivered by catering companies. A full-scale commercial kitchen will process and cook food for wholesale to public/private clients and individual customers through the canteen. The Kitchen Food Hub plays a crucial role in a paradigm shift: from the commodification of food and the daily life of children and adult citizens to the broader concept of the "common good" that serves the urban and agricultural community. This kitchen will accommodate facilities to transform raw ingredients into meals including meat and vegetable processing areas. Therefore, it can enhance local products without requiring intermediate steps and pre-processed operations and ingredients.

The aggregation, processing and distribution centre serves as the terminal destination along the various supply chains that lead to the Kitchen Food Hub from farms and packing houses in the surrounding community. Arriving products may be aggregated, washed, sorted and graded here if a packing house does not provide this service before arrival. Once the food is ready for wholesale distribution, it is stored in cold and dry storage units before being loaded onto refrigerated trucks for delivery to various clients such as institutions, schools and restaurants.

The kitchen incubator provides start-up food processing entrepreneurs with commissary kitchen space that meets health and safety regulations. Food entrepreneurs will have wholesale accounts with the aggregation, processing and distribution centre to directly source produce, meat, eggs, honey, dairy and other local farm products. This space also functions as an additional teaching/event kitchen to support taste and nutrition education, product promotion and destination itineraries related to cooking classes, demonstrations and community or private events.

The canteen provides citizens with an educational eating/dining experience where menu and product development are transparent and seasonal. Made from fresh and sustainably produced ingredients, the canteen offers a balanced meal option for all needs of a diverse clientele. This space welcomes citizens to participate and celebrate the importance of good food and promote well-being in daily life. The canteen focuses on involving schools to engage children in understanding how the food system works and how they can live with a balanced, sustainable diet while offering adult urbanites a good meal option daily.

The farmers market and meeting room is where all actors involved in the food supply chain, "from seed to stomach", come together. While food producers and processors can engage with new customers in the market, the meeting room is an educational hub integrating agricultural research, training for farmers and food professionals, and activities that connect the urban with the rural domain. The meeting room is available for private and public use, classes and thematic courses. A farmers market is organised once a week, involving food producers and processors that are part of the cluster supported by the Kitchen Food Hub, offering businesses (B2B) and citizens (B2C) the opportunity to meet and purchase directly from the source of production.

These KFHs should be able to produce 2/3000 meals per shift up to 4/6.000 if they will also operate in providing evening meals instead of only lunch and buy smaller batches of raw ingredients, allowing small local producers to access this market.

Moreover, the concept is modular and scalable; it will be possible to have a central facility (KFH) and many different production units within the city/territory in which it is implemented to produce meals as close as possible to places of consumption, such as school canteens, hospitals, nursing home, prisons. The KFH become an essential asset for cities in their food resilience, becoming part of their urban food system.

Why we need a new paradigm

The current food system faces significant challenges, such as climate change, biodiversity loss, food insecurity and health problems, is mainly based on a linear model of production, consumption and disposal, which is unsustainable in the long run. A paradigm shift is urgently required to face the food dilemma and address these challenges.

A circular model is needed to shift to a more sustainable food system. This model is based on the principles of reduce, reuse and recycle, emphasising the need to minimise waste, optimise resource use and enhance resilience.

One of the critical shifts that is needed is a move towards more sustainable and regenerative farming practices. This includes practices such as small-scale fishery, agroforestry, agroecology and organic agriculture which can help to reduce the use of synthetic fertilisers and pesticides, improve soil and water health, and increase biodiversity.

Another significant shift is towards more equitable and just food systems. This includes ensuring that farmers and fisherman are paid fair prices for their products, workers in the food system are paid living wages, and everyone has access to healthy and affordable food.

Finally, there is a need for a shift towards more plant-based diets. The production of meat and dairy products is resource-intensive and significantly impacts the environment. By shifting towards more plant-based diets, we can reduce the demand for animal products and create a more sustainable food system.

Overall, a paradigm shift in the food system requires a holistic and integrated approach that considers social, cultural, environmental and economic factors. This shift will require the cooperation and collaboration of all stakeholders, including governments, industry, farmers, consumers and civil society.

Conjunctural adjustments, partial solutions or extraordinary interventions are not enough; we need to intervene progressively in the whole system adopting an active stepwise policy transition to integrate sustainable food system thinking into all existing EU policies related to food & health.

The health and well-being challenge

Europe is experiencing a significant demographic shift towards an aging population. This demographic shift results from declining birth rates and increased life expectancy, leading to a more significant proportion of the population over 65. According to Eurostat, the statistical office of the European Union, the share of the population aged 65 and over is projected to increase from 20% in 2020 to 31% in 2100 (Eurostat, 2019).

While life expectancy has been increasing in many high-income countries, recent studies (Spiers et al., 2021) have shown that there has also been an expansion of disability in later life. This means that people are living longer but are also experiencing more years of poor health and disability, which has significant implications for the sustainability of health and care systems.

Food service for nursing homes should prioritise meeting the nutritional needs tailored to meet the dietary needs of residents while also providing tasty, visually appealing and safe meals. Furthermore, providing culturally appropriate meals can help residents feel more comfortable and at home. This service may be provided in several ways, including on-site kitchens, contracted catering services or a combination of both.

Therefore, we will need to give the utmost attention to food services inside nursing homes; it is also critical that for all younger generations, the greatest attention should be given to service levels and the quality of meals served inside the school and university canteen.

We need a new commitment of teachers, caretakers, parents, children, young people, caterers and suppliers, working in partnership at the national and local level to ensure that European young people enjoy a healthier, thriving, sustainable and resilient food future.

In 2019 Commission President Ursula von der Leyen announced in her Political Guidelines the creation of a European Child Guarantee intending to ensure that every child in Europe at risk of poverty or social exclusion has access to the most basic of rights like healthcare and education.

On 24 March 2021, after consultations with key stakeholders, including the children themselves, European Commission adopted a proposal for Council Recommendation establishing a European Child Guarantee; the objective of the European Child Guarantee is to prevent and combat social exclusion by guaranteeing adequate access to children in need to a set of key services, including school-based activities and at least one healthy meal each school day.

Several international agreements and national policies recognise children's rights to healthy food in school canteens. Some examples include:

- The United Nations Convention on the Rights of the Child recognises the right of every child to the highest attainable standard of health, including access to nutritious food and clean water
- The Global Strategy for Women's, Children's and Adolescents' Health, which calls for the promotion of healthy diets for children and the reduction of malnutrition

- The WHO Global Action Plan for the Prevention and Control of Noncommunicable Diseases recommends implementing policies and interventions to promote healthy diets, including in school settings
- In France, the Egalim Law of 2018 mandates that school canteens serve at least 50% of organic, local or sustainable food and that the menu must meet specific nutritional guidelines

Ensuring that children have access to healthy food in school canteens is vital for their physical and mental well-being and academic success. It can also contribute to developing healthy eating habits and promote the consumption of nutritious food.

Environmental and social challenges

Food public procurement faces several environmental and social challenges that must be addressed to promote sustainability and social responsibility.

The environmental impact can be significant due to the large quantities of food purchased and consumed. This impact is driven by food production, processing, packaging, transportation and waste disposal, which contributes to greenhouse gas emissions and environmental degradation. This waste also represents a significant financial loss for public institutions; furthermore, shortening supply chains can significantly reduce these impacts.

The food service industry and, consequently, the food public procurement can also have significant impacts on the following:

- Animal welfare depends on how animals are raised and treated for food production
- Cultural identity, food is often closely tied to and can play an essential role in preserving cultural heritage and traditions by offering a variety of cuisines and incorporating cultural practices into food service operations
- Employment provides job opportunities for individuals at all skill levels. However, low wages and lack of benefits in some industry sectors can contribute to economic inequality and phenomena like labour mismatch
- School canteen and other food service premises can contribute to the social fabric of a community by serving as gathering places and contributing to the local economy

Addressing these challenges requires a commitment to sustainability and social responsibility from public institutions.

Solutions include promoting plant-based diets, reducing food waste, improving food labelling and transparency, promoting sustainable agriculture and fishery practices, improving labour practices and promoting animal welfare.

The integration of recommendations that promote specific food practices and choices while also addressing the ecological dimensions of nutrition would represent an efficient strategy to contribute to the reshaping of the current food systems towards the provision of health-promoting products that are less impacting on the environment (Rossi et al., 2023).

Public health prevention

Public health prevention is a crucial component of promoting the health and well-being of individuals and communities. It involves a range of interventions and strategies that address the social, economic and environmental determinants of health.

Parents in Europe want their children to grow up healthy, happy and ready to succeed in life, but childhood overweight and obesity undermine children's health.

Overweight and obesity are threatening our children's future and the future of our cities, which looks to their children as the next generation of citizens and leaders. If our children are not healthy, then our society will not flourish. Overweight and obesity and other non-communicable diseases (NCD) also threaten the financial sustainability of our different national healthcare systems.

A diet based on whole foods, as opposed to processed foods, can play an essential role in health prevention. A recent study suggests that replacing processed and ultra-processed foods and drinks with an equal amount of minimally processed foods might reduce the risk of various cancer types. (Kliemann et al., 2023)

It is a significant opportunity to rise to these health, social and environmental challenges, but to achieve any goals on this issue we need to educate and inform consumers, and especially youth.

Food literacy and food education

In an age teeming with nutritional paradoxes and dietary conundrums, instilling food education and fostering food literacy have emerged as pivotal components of a comprehensive approach to health and well-being.

These twin pillars are instrumental in empowering individuals, particularly the younger generation, to navigate the intricate food landscape, make informed decisions and cultivate sustainable eating habits.

Food education extends beyond the mere identification of fruits and vegetables. It encompasses a broad spectrum of knowledge areas, including nutritional science, food origins, preparation techniques and cultural culinary traditions. Integrating food education into school curricula and community programmes plants the seeds of knowledge, enabling children and adults alike to comprehend the correlation between food, health and the environment.

School-based garden programmes and cooking classes exemplify practical avenues for food education, offering hands-on experiences that demystify the journey of food from farm to fork. These initiatives not only enhance understanding but also foster an appreciation for the diversity and cultural richness embedded in culinary traditions worldwide.

While food education lays the foundation, food literacy equips individuals with the critical thinking skills necessary to interpret and apply this knowledge. It encompasses the ability to interpret food labels, assess marketing claims, manage food budgets and make ethical and sustainable food choices.

In the face of an ever-expanding array of food products and a bombardment of often conflicting dietary advice, food literacy serves as a compass, guiding individuals through the maze of choices. It fosters a discerning approach to consumption, enabling individuals to sift through misinformation and make decisions aligned with their health goals and values.

The symbiotic relationship between food education and food literacy is the linchpin for cultivating a society characterised by health-conscious, informed and responsible consumers. Together, they unravel the complexities of modern food systems, promote healthy eating habits and contribute to the mitigation of diet-related health issues such as obesity and diabetes.

Investing in food education and nurturing food literacy are not mere academic exercises; they are essential ingredients in the recipe for a healthier, more equitable and sustainable future. By fostering an environment where knowledge about food is accessible and valued, we are not just feeding minds but also nourishing communities and sowing the seeds for a flourishing global food landscape.

This is why Food education programmes should also be part of any public tender for schools meal services.

A "whole school food approach", introduced into the EU-funded project "School Food 4 Change",[1] is a comprehensive approach to school food that involves all aspects of the school environment, including food procurement, preparation and service, food education and policy development.

A whole school food approach aims to create a school food environment that promotes health, well-being and sustainability.

The whole school food approach (WSFA) is a holistic concept that offers pupils healthier and more sustainable food choices in the long term. It is not just about what food they are served in schools but also about where the food comes from and how production impacts our environment and our health. Putting the WSFA into practice requires teamwork from school staff, cooks, schoolchildren and all other stakeholders. This concept can also be extended, albeit in a slightly different form, to high schools and universities, where pupils become students and the next actors in the society of the future.

Here are some key components of a whole-school food approach:

- Food Procurement: A whole school food approach emphasises the importance of sourcing healthy, sustainable and locally produced food whenever possible. This includes working with local farmers and food producers to supply fresh, seasonal produce and choosing foods that are minimally processed and free of additives and preservatives
- Food Preparation and Service: Schools that adopt a whole-school food approach prioritise preparing and serving fresh, healthy meals that meet nutritional standards. This may involve cooking from scratch, avoiding highly processed and fast foods, and using cooking methods that preserve the nutritional value of food
- Food Education: A whole school food approach also includes opportunities for pupils and students to learn about healthy eating and food systems. This can take the form of nutrition education in the classroom, cooking classes and hands-on learning experiences such as school gardens and farm-to-school programmes
- Policy Development: Schools that adopt a whole-school food approach may develop policies that support healthy and sustainable food practices. This could

include policies around food procurement, meal standards, food waste reduction and food marketing in schools

By adopting a whole-school food approach, schools can create a healthy and sustainable food environment that supports the health and well-being of pupils, students, staff and the wider community.

Pupils and students, particularly in school settings, can also be considered "food evangelists" as they have the potential to influence their peers and communities towards more sustainable and healthy food choices (Ekasasi, 2005).

Through their actions and advocacy, the youth can help create a more sustainable and healthy food system within their schools and communities. Their voices and actions can help to drive change toward a more just, equitable and sustainable food system for all.

Note

1 https://schoolfood4change.eu/blog/project/whole-school-food-approach/.

References

Ekasasi, S. R. (2005). The Role of Children in Family Decision Making a Theoretical Review. *Jurnal Siasat Bisnis*. https://journal.uii.ac.id/JSB/article/view/977

Eurostat. (2019, 10 July). *The EU's Population Projected up to 2100*. https://ec.europa.eu/eurostat/web/products-eurostat-news/-/DDN-20190710-1

Kliemann, N., Rauber, F., Bertazzi Levy, R., Viallon, V., Vamos, E. P., Cordova, R., Freisling, H., Casagrande, C., Nicolas, G., Aune, D., Tsilidis, K. K., Heath, A., Schulze, M. B., Jannasch, F., Srour, B., Kaaks, R., Rodriguez-Barranco, M., Tagliabue, G., Agudo, A., Panico, S., . . . Huybrechts, I. (2023). Food Processing and Cancer Risk in Europe: Results From the Prospective EPIC Cohort Study. *The Lancet. Planetary Health* 7(3), e219–e232. https://doi.org/10.1016/S2542-5196(23)00021-9

Pollan, M. (2007). *The Omnivore's Dilemma: A Natural History of Four Meals*. New York: Penguin,

Rossi, L., Ferrari, M., & Ghiselli, A. (2023). The Alignment of Recommendations of Dietary Guidelines with Sustainability Aspects: Lessons Learned from Italy's Example and Proposals for Future Development. *Nutrients* 15, 542. https://doi.org/10.3390/nu15030542

Spiers, G. F., Kunonga, T. P., Beyer F., et al. (2021). Trends in Health Expectancies: A Systematic Review of International Evidence. *BMJ Open* 11, e045567. https://doi.org/10.1136/bmjopen-2020-045567

Index

Taylor & Francis Group
an informa business

Taylor & Francis eBooks

www.taylorfrancis.com

A single destination for eBooks from Taylor & Francis with increased functionality and an improved user experience to meet the needs of our customers.

90,000+ eBooks of award-winning academic content in Humanities, Social Science, Science, Technology, Engineering, and Medical written by a global network of editors and authors.

TAYLOR & FRANCIS EBOOKS OFFERS:

A streamlined experience for our library customers

A single point of discovery for all of our eBook content

Improved search and discovery of content at both book and chapter level

REQUEST A FREE TRIAL

support@taylorfrancis.com

Routledge
Taylor & Francis Group

CRC Press
Taylor & Francis Group

Printed in the United States
by Baker & Taylor Publisher Services